USE OF IOT
Internet of Things

USE OF IOT
Internet of Things

Editors
DR. ARPIT JAIN
PROF. (DR.) RATNA RAJA KUMAR JAMBI
PROF. (DR.) B. K. SARKAR
PROF. (DR.) VIVEK KUMAR

NAMYA PRESS.com

Published by

Namya Press

USA: 10685-B Hazlehurst Dr. # 31013 Houston, Texas - 77043
Email: namyapress@gmail.com
Website: https://namyapress.com

Edition	:	First Published in 2022 (Namya Press)
Title	:	USE of IoT: Internet of Things
Editors	:	Dr. Arpit Jain, Prof. (Dr.) Ratna Raja Kumar Jambi, Prof. (Dr.) B. K. Sarkar, Prof. (Dr.) Vivek Kumar
ISBN	:	978-1-956861-10-5
Copyright	:	© Editors 2022. All rights reserved

Preface

It's with great happiness that I would like to acknowledge a great deal of people that get helped me extremely through the entire difficult, challenging, but a rewarding and interesting path towards some sort of Edited Book Without having their help and support, none of this work could have been possible.

Prof. (Dr.) B.K Sarkar: I am forever indebted to my father **Mr. Ichchha Ram Singh**, My mother **Ms. Ram Sohari Devi** and also the rest of my children **Mr. Pawan Kumar Singh** and **Miss Parinidhi Singh**. My wife **Prof. (Dr.) Reena Singh** is a huge great source of inspiration in my experience. None of this will have been achieved without her really like, support and steady encouragement and Also Support My All National and International Friends.

My parents' prayers have always accompanied me. Their love keeps me ambulatory. I am grateful to all of them.

Dr. Arpit Jain: I am forever indebted to my father **Mr. J. B. Jain**, My mother **Ms. Manju Jain** and also the rest of my children **Vaani Jain** and **Vrasti Jain**. My wife **Garima Jain** is a huge great source of inspiration in my experience. None of this will have been achieved without her really like, support and steady encouragement.

My parents' prayers have always accompanied me. Their love keeps me ambulatory. I am grateful to all of them.

Contents

Chapter - 01

IoT - Based Digital Marketing

Prof. (Dr.) Kirti Mahajan,

[1]Professor & Director (International Affairs),
Bharati Vidyapeeth (Deemed to be University), Institute of Management and Entrepreneurship Development, Pune, Bharati Bhawan, LBS Marg, Pune 411030

Mr. Sunil Sahu

[2]Entrepreneur, Author of self learning Books, Open Source Architect, and Managing Director of NCS Pvt Ltd, has been working in IT industry since last 22 years on J2EE, Cloud, DevOPS, Data Science technologies.

ABSTRACT

The developing peculiarity of the Internet of Things (IoT), which is that any thing equipped for being associated with the Internet will be, presents a remarkable chance for organizations. Utilizing a broad writing audit, the flow research analyzes the critical change in showcasing techniques that need to happen to focus on the millennial age of as they embrace IoT. Most examination characterizes the Millennial age as those brought into the world from the mid 1980s to the mid 2000s. As people in this age become older, there are two social ramifications: 1) their acknowledgment of innovation proposes they will rush to embrace IoT, and 2) their developing buying power and purchaser conduct make them an optimal objective for advertisers. Recent college grads who take on IoT offer their information all the more eagerly to advertisers and firms, which makes it simpler for advertisers to exactly gather information and target clients more. Furthermore, IoT gadgets will empower different stages for content promoting that are fundamentally unique in relation to and more successful than a 30 second TV plug or an advanced pennant notice. Advertising messages will be more customized, modified, and focused on to likely clients than any time in recent memory. Besides, IoT offers limitless imagination for content creation as well as designated conveyance of content, instead of conventional publicizing roads.

BACKGROUND OF THE INVENTION

IoT is still considered by many in its infancy, though its applications are extremely advanced. To understand a background of IoT, one must first understand the evolution of the Web. The Internet is undoubtedly the "largest transformable information construct."

Between 2

Smartphones, computers, and other devices, much of today's society is embedded in and powered by the Internet. However, the evolutionary pattern of the Internet during the past two decades is distinct. Scholars define the various stages of web development as "Web 1.0 as a tool for cognition, 2.0 as a medium for human communication, and 3.0 as networked digital technology 3 that supports human cooperation." Simultaneous to Web 1.0, IoT was birthed as MachinetoMachine communication (M2M) around 1989. M2M technology is nothing radically new in the way that the Internet is not. For example, airplane communication devices are constant M2M feeds that autonomously send information to one another from ground towers to airborne planes.

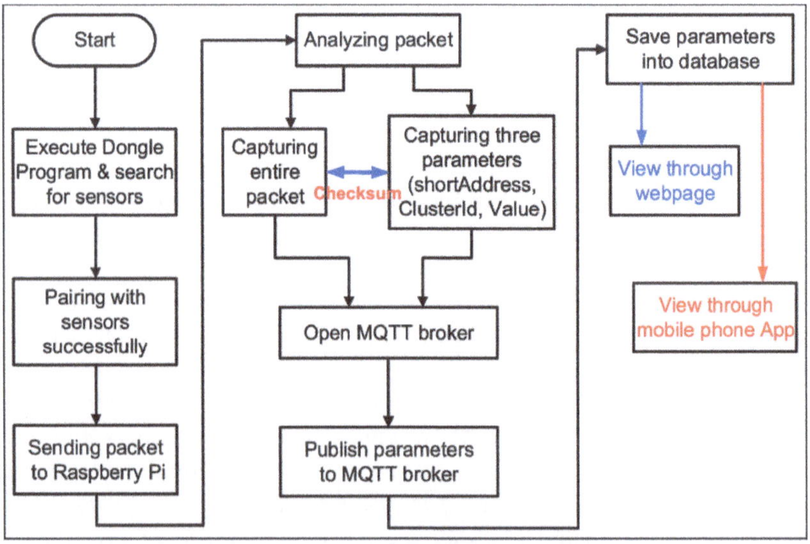

Fig. 1: IoT- Based Digital Marketing

OBJECTIVES OF THE INVENTION

A clear definition of the product is necessary for consumers to understand what exactly is being offered, and IoT allows for highly customized products. When a consumer is able to enjoy a customized product, that product is able to provide truly unique experiences; something *that is highly valued by Millennials. he placement of a product is typically its channel of distribution, such as whether the product can be purchased in physical stores or online marketplaces.*

Within a store, placement can be further divided into the physical location such as the checkout line, end aisle displays or back of the store. Placement also applies to marketing messaging, such as advertisements on billboards, television, radio, or digital banners. IoT changes the way consumers interact with marketing messages because the IoT devices enable consumers to be constantly connected to the firm, thus placement is anywhere the consumer is located.

The price of a product needs to be clearly determined as it communicates value to consumers. Like message placement, IoT changes this definition. IoT connected devices allow for dynamic pricing of items based on actual behavior, as opposed to somewhat traditional segmentation based on consumer demographics.

For example, Progressive Auto Insurance has a device called the Snapshot that allows the company to collect actual driver information. This allows Progressive to dynamically price insurance based on actual driver behaviors as opposed to blanket pricing across demographics such as age or gender.

SUMMARY OF THE INVENTION

Tracking Behavior Utilizing products embedded with sensors allows companies to easily monitor the behavior of persons, things, or data through space and time. IoT enabled devices are inherently sensordriven and collect data from numerous touch points depending on the architecture of the device including movement, physical activity, communication patterns, and more. Companies that implement IoT in their marketing plans allow for their "business models [to] be finetuned to take advantage of this behavioral data."

Enhanced Situational Awareness Aggregating several of these sensors collecting data from different touch points implies that these devices can become so advanced as to have enhanced situational awareness. Achieving realtime information of the physical environment and how consumers are interacting with it allows for decision makers to understand an environment before taking action. This allows for marketers to triangulate behavior and get a deeper, contextualized understanding of the patterns of consumer behaviors unlike ever before.

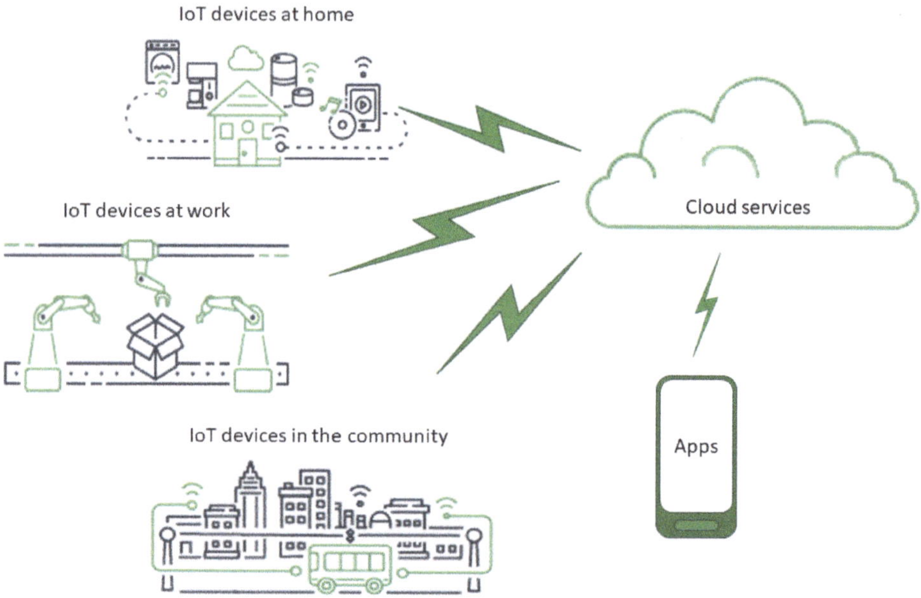

Fig. 2: IoT- Based Digital Marketing

Sensor Driven Analytics Spilotro

11 Taking all of this data into consideration, analytics can provide more digestible content for marketers. Sensor driven analytics can assist human decision making through deep analysis and data visualization software. Contextualizing the data leads to marketing automation and control unlike ever before, drastically reducing the burden on marketers to segment and find their audience. This allows marketing teams to focus on creating effective content for a few smaller segments rather than chasing a larger target market and hoping content is effective.

DESCRIPTION OF THE INVENTION

More Money, More Targeting Another reason to target Millennials is their growing purchasing power. According to Merrill Lynch,

Millennials' purchasing power is projected to reach $8.1 Trillion by 2025, as 21 Millennials will use their disposable income to purchase new technology and other luxury (nonessential) items. For example, younger millennials are more likely to purchase bigticket items such as homes, whereas older millennials are more likely to rent. 22

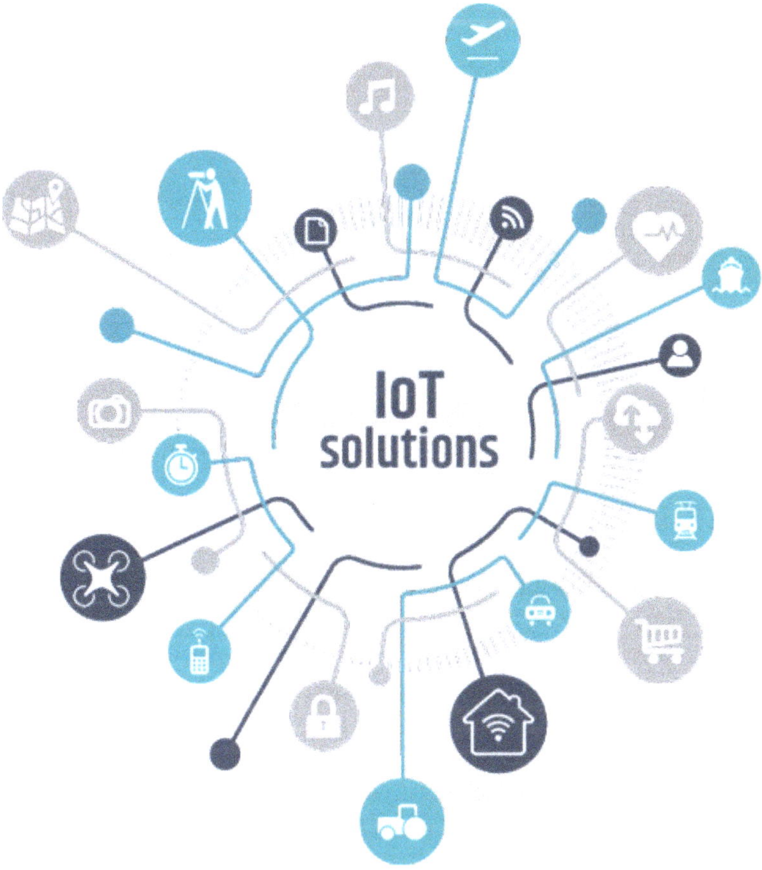

Fig. 3: IoT- Based Digital Marketing

Similarly, they may be more willing to purchase a technologyenabled car (i.e., smart car) that firms will be able to equip with IoT features. As such, firms need to be cognizant of the divide within this generation and should alter their products and services in an appropriate manner.

Content Marketing Spilotro 14 In an effort to solve this problem, marketers have shifted the focus from traditional marketing efforts to content marketing. Content marketing focuses on "*creating and distributing relevant and valuable content to attract, acquire, and engage a target audience with the objective of driving profitable customer action.*" 26 Millennials are savvy content consumers.

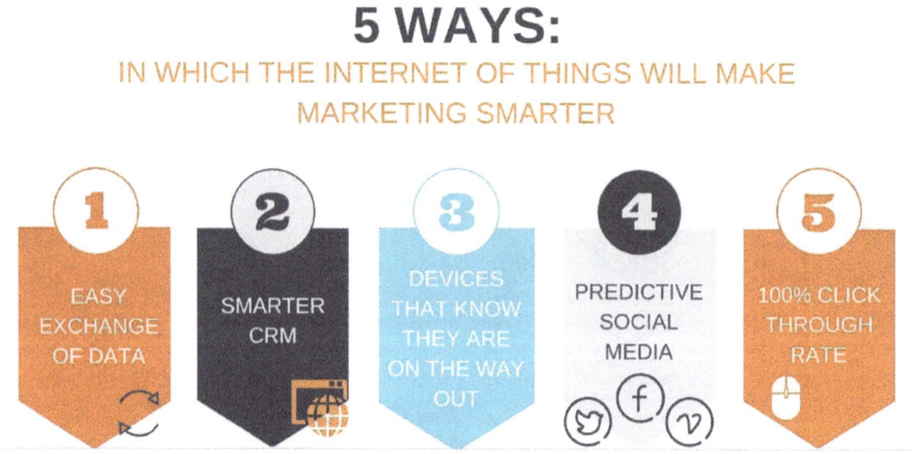

Fig. 4: IoT- Based Digital Marketing

As previously mentioned, technological advances mean that brands have many different avenues by which to reach their audience, but that can be a doubleedged sword. Millennials are keenly aware of how to leverage their attention, and because they are the highesteducated generation, they expect brands to be transparent in their marketing efforts. To effectively target millennials "requires brands to tell stories, create shareworthy content, and foster an experience around a 27 product/service." As such, companies should not attempt to fool Millennials as it will not only backfire in the short term of losing a sale, but deceptive practices (regardless of legitimacy) can damage Millennials' relationship with a brand, sometimes irreversibly.

CONSUMER PRODUCTS

IoT marketing applications are not just limited to technological partnerships. Recently, one company called Appirio experimented with a vending machine powered by tweets. To get a product from the vending machine, users had to interact with it on Twitter. This provides a 34 unique user experience geared toward Millennial behaviors of sharing on social media. Another example is Johnnie Walker whiskey. In 2015, parent company Diageo decided to connect 100,000 bottles to the internet so they could be used to send a personalised, onetoone digital video messages to dad in honor of Father's Day. Doing this helped Diageo involve Millennials 1) in the process of marketing and 2) generating the content. As a result, Diageo saw a 72% sales uplift leading up to Father's Day, and the cost of the campaign was repaid five times 35 over. This campaign also helped to portray Johnnie Walker as innovative, positioned whiskey as the perfect Father's day gift, and created a unique user experience both for the giver and the recipient of the product.

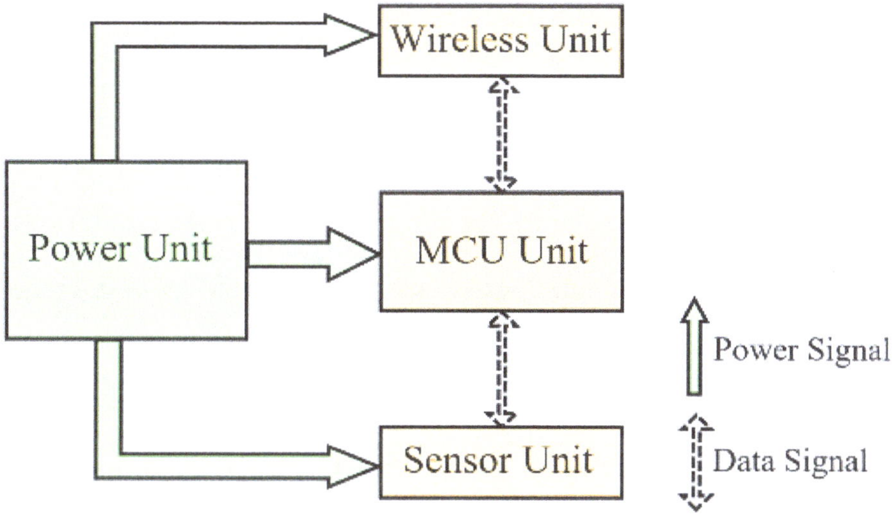

Fig. 5: IoT- Based Digital Marketing, Unit

Health and Fitness Health conscious consumers could purchase an IoT connected cooking pan that analyzes the nutritional value of the food they're cooking. It could be programmed to automatically send that data to a food log app on their smartphone such as MyFitnessPal (powered by Under Armour). This would provide Under Armour with specific demographic information and insight into key behaviors of their consumers. For instance, based on eating habits, Under Armour could discern which of their customers are serious athletes or purchase clothing for "athleisure."

With that information, Under Armour could target more effectively with dynamic marketing messages tailored to different segments. Similarly, connecting that information with mobile apps for grocery stores would be beneficial. Grocers could also use this information to see what customers are actually cooking with their products. They could also send push notifications based on expiration dates to alert consumers when their food is close to spoiling or running out.

For a healthoriented generation, this too would be an application we might see sooner rather than later. Additionally, this would help to ensure that grocery stores' inventories are not depleted because they will know ahead of time what products are being consumed and when a customer is likely to make a repeat purchase.

Real Estate

Spilotro 21 Younger millennials are sooner apt to purchase a home in their lifetime than older millennials, so there is an opportunity to use IoT devices in this industry. Real estate agents could use IoT devices to quickly collect information of those attending an open house so as to follow up with similar listings. Additionally, they could target consumers for other homes by coupling previously collected data with geolocation services to send push notifications to their phone when they are in an area with similar listings that are listed by that agent.

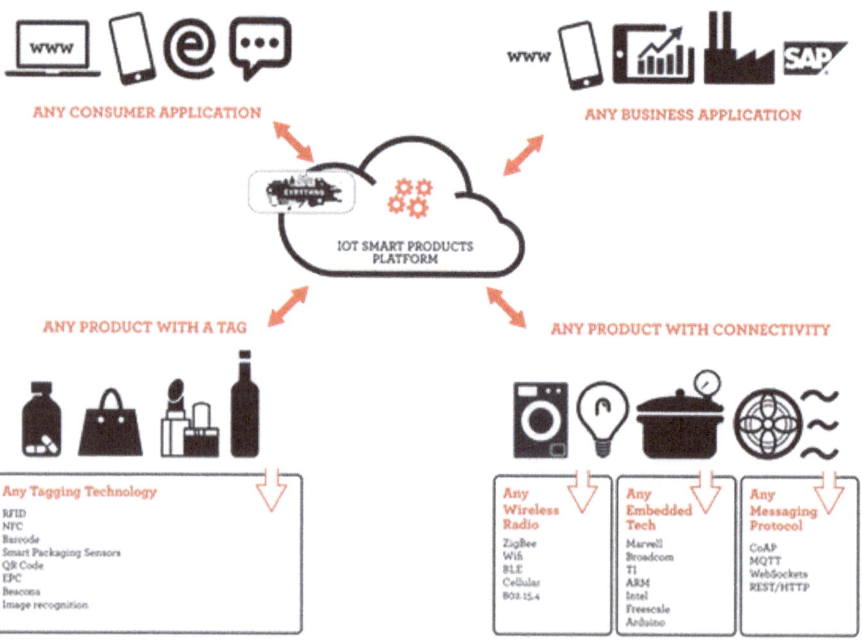

Fig. 6: IoT- Based Digital Internet Marketing.

Result

A clear definition of the product is necessary for consumers to understand what exactly is being offered, and IoT allows for highly customized products. When a consumer is able to enjoy a customized product, that product is able to provide truly unique experiences; something *that is highly valued by Millennials. The placement of a product is typically its channel of distribution, such as whether the product can be purchased in physical stores or online marketplaces.*

Within a store, placement can be further divided into the physical location such as the checkout line, end aisle displays or back of the store. Placement also applies to marketing messaging, such as advertisements on billboards, television, radio, or digital banners. IoT changes the way consumers interact with marketing messages because the IoT devices enable consumers to be constantly connected to the firm, thus placement is anywhere the consumer is located.

The price of a product needs to be clearly determined as it communicates value to consumers. Like message placement, IoT changes this definition. IoT connected devices allow for dynamic pricing of items based on actual behavior, as opposed to somewhat traditional segmentation based on consumer demographics.

For example, Progressive Auto Insurance has a device called the Snapshot that allows the company to collect actual driver information. This allows Progressive to dynamically price insurance based on actual driver behaviors as opposed to blanket pricing across demographics such as age or gender.

REFERENCES

[1] Balaji, M.; Roy, S.K. Value co-creation with Internet of things technology in the retail industry. J. Mark. Manag. 2022, 33, 7–31. [CrossRef]

[2] Ikävalko, H.; Turkama, P.; Smedlund, A. Value Creation in the Internet of Things: Mapping Business Models and Ecosystem Roles. Technol. Innov. Manag. Rev. 2020, 8, 5–15. [CrossRef]

[3] Dijkman, R.M.; Sprenkels, B.; Peeters, T.T.; Janssen, A. Business models for the Internet of Things. Int. J. Inf. Manag. 2019, 35, 672–678. [CrossRef]

[4] Ashton, K. That 'Internet of Things' thing. RFID J. 2019, 22, 97–114.

[5] Ng, I.C.; Wakenshaw, S.Y. The Internet-of-Things: Review and research directions. Int. J. Res. Mark. 2017, 34, 3–21. [CrossRef]

[6] Woodside, A.G.; Sood, S. Vignettes in the two-step arrival of the internet of things and its reshaping of marketing management's service-dominant logic. J. Mark. Manag. 2016, 33, 98–110. [CrossRef]

[7] Mosaic. Mosaic Consumer Classification. Available online: https://www.experian.co.uk/marketing-services/ products/mosaic-uk.html (accessed on 11 December 2018).

[8] Bi, Z.; Cochran, D. Big data analytics with applications. J. Manag. Anal. 2014, 1, 249–265. [CrossRef]

[9] DPA. UK Data Protection Act 1998. Available online: http://www.legislation.gov.uk/ukpga/1998/29/contents (accessed on 11 December 2018).

CITATIONS

- Jain, A., Kumar, A. and Sharma, S., 2015. Comparative Design and Analysis of Mesh, Torus and Ring NoC. Procedia Computer Science, 48, pp.330-337.
- Jain, A., Dwivedi, R., Kumar, A. and Sharma, S., 2017. Scalable design and synthesis of 3D mesh network on chip. In Proceeding of International Conference on Intelligent Communication, Control and Devices (pp. 661-666). Springer, Singapore.
- Jain, A., Gahlot, A.K., Dwivedi, R., Kumar, A. and Sharma, S.K., 2018. Fat Tree NoC Design and Synthesis. In Intelligent Communication, Control and Devices (pp. 1749-1756). Springer, Singapore.
- Jain, A., Dwivedi, R., Kumar, A. and Sharma, S., 2017. Scalable design and synthesis of 3D mesh network on chip. In Proceeding of International Conference on Intelligent Communication, Control and Devices (pp. 661-666). Springer, Singapore.
- Sharma, S.K., Jain, A., Gupta, K., Prasad, D. and Singh, V., 2019. An internal schematic view and simulation of major diagonal mesh network-on-chip. Journal of Computational and Theoretical Nanoscience, 16(10), pp. 4412-4417.

CHAPTER - 02

Medical Data Transmission Using IoT

Neeraj Milind Shahane

CEO, D-Tech, Philter Co, U1/26 Arnott street Clayton vic 3168 Email : shahaneneeraj00@gmail.com

Pitty Nagarjuna,

Senior Scientific Officer, Indian Institute of Science, Bengalore 560012 nagarjuna@iisc.ac.in

Prof. (Dr.) Vivek Kumar

Vice Chancellor, Quantum University, Roorkee, UK, India. vicechancellor@quantumuniversity.edu.in

Prof. (Dr.) B.K. Sarkar

MES's, Pillai HOC College of Engineering &Technology, Rasayani Taluka Panvel, Dist, Navi Mumbai, Maharashtra 410207, India. dr.bksarkar2003@yahoo.in

ABSTRACT

Because of the huge progression of the web of things (IoT) in the medical care area, the security and the respectability of the clinical information turned out to be enormous difficulties for medical services administrations applications. This paper proposes a half breed security model for getting the analytic text information in clinical pictures. The proposed model is created through coordinating either 2D Discrete Wavelet Transform 1 Level (2D-DWT-1L) or 2D Discrete Wavelet Transform 2 Level (2D-DWT-2L) steganography procedure with a proposed mixture encryption conspire. The proposed mixture encryption outline is assembled utilizing a blend of Advanced Encryption Standard (AES), and Rivest, Shamir, and Adleman (RSA) calculations. The proposed model beginnings by scrambling the restricted information; then, at that point, it conceals the outcome in a cover picture utilizing 2D-DWT-1L or 2D-DWT-2L. Both variety and dim scale pictures are utilized as cover pictures to hide different text sizes. The exhibition of the proposed framework was assessed in view of six measurable boundaries; the Peak Signal to Noise Ratio (PSNR), Mean Square

Error (MSE), Bit Error Rate (BER), Structural Similarity (SSIM), Structural Content (SC), and Correlation. The PSNR values were moderately shifted from 50.59 to 57.44 in the event of variety pictures and from 50.52 to 56.09 with the dark scale pictures. MSE values differed from 0.12 to 0.57for the variety pictures and from 0.14 to 0.57 for the dark scale pictures. BER values were zero for the two pictures, while SSIM, SC and Correlation values were ones for the two pictures. Contrasted with the best in class techniques, the proposed model demonstrated its capacity to conceal the classified patient's information into a sent cover picture with high indistinctness, limit, and negligible crumbling in the got stego-picture.

Index Terms: *Cryptography, DWT-1level, DWT-2level, Encryption, Healthcare Services, Internet of Things, Medical Images, Steganography*

INTRODUCTION

IoT creates an integrated communication environment of interconnected devices and platforms by engaging both virtual and physical world together [1]. With the advent of remote digital healthcare based IoT systems, the transmission of medical data becomes a daily routine. Therefore, it is necessary to develop an efficient model to ensure the security and integrity of the patient's diagnostic data transmitted and received from IoT environment [2-8]. This goal is carried out using steganography techniques and system encryption algorithms together to hide digital information in an image [9-6]. Cryptography is another term for data encryption [1][7].

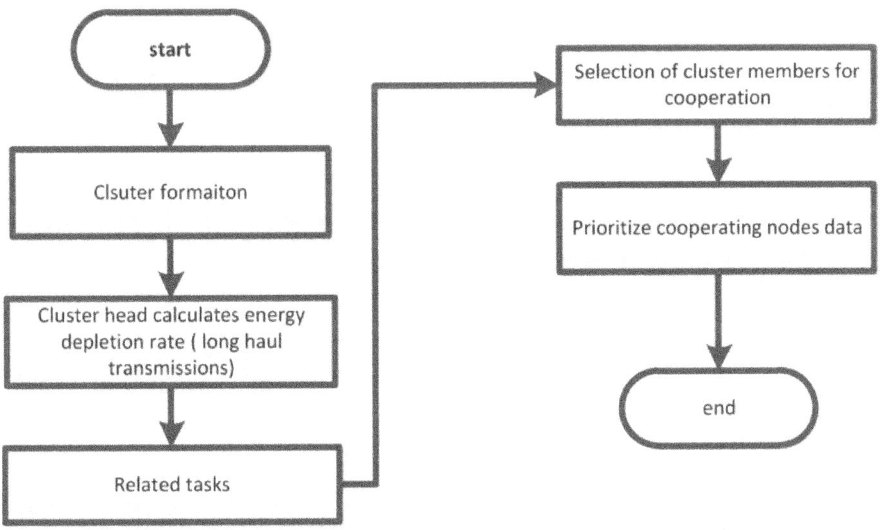

Fig. 1: Medical Data Transmission using IoT

Encryption cryptography is the process of encoding messages in a way that hackers cannot read it, but that can be authorized personnel. The two main algorithms used for data encryption in this work are the Advanced Encryption Standard (AES) and the Rivest-Shamir-Adleman (RSA) algorithm [1][8]. AES is a symmetric cipher where the same key is used on both sides [1][9].

It has a fixed message block size of 128 bits of text (plain or cipher), and keys of length 128,192, or 256 bits. When longer messages are sent, they are divided into 128-bit blocks. Apparently, longer keys make the cipher more difficult to break, but also enforce a longer encrypt and decrypt process. On the contrary, the RSA is a public key algorithm, which widely used in business and personal communication sectors [2][0]. It has the advantage of having a variable key size ranging from (2-2048) bits.

THE PROPOSED MODEL

This paper proposes a healthcare security model for securing a medical data transmission in IoT environments. The proposed model composes of four continuous processes: (1) The confidential patient's data is encrypted using a proposed hybrid encryption scheme that is developed from both AES and RSA encryption algorithms. (2) The encrypted data is being concealed in a cover image using either 2D-DWT-1L or 2D-DWT-2L and produces a stego-image. (3) The embedded data is extracted. (4) The extracted data is decrypted to retrieve the original data. Fig. 1 shows the general framework of our proposed model for securing the medical data transmission at both the source's and the destination's sides.

Data Encryption Scheme

The proposed model implements the cryptographic scheme. The cryptographic scheme Ĉ ൭ന, ଢ, C, S, T is composed of encryption and decryption processes. Throughout the encryption process, the plain text T is divided into odd part T୨ and even parts Ţ୭. The AES is used to encrypt T୨ using a secret public key s. The RSA is used to encrypt Ţ୭ using a secret public key m.

The private key x that used in the decryption process at the receiver side is encrypted using AES algorithm and sent to the receiver in an encrypted form to increase the security level. The encryption process can be mathematically modeled as given in the following equations below.

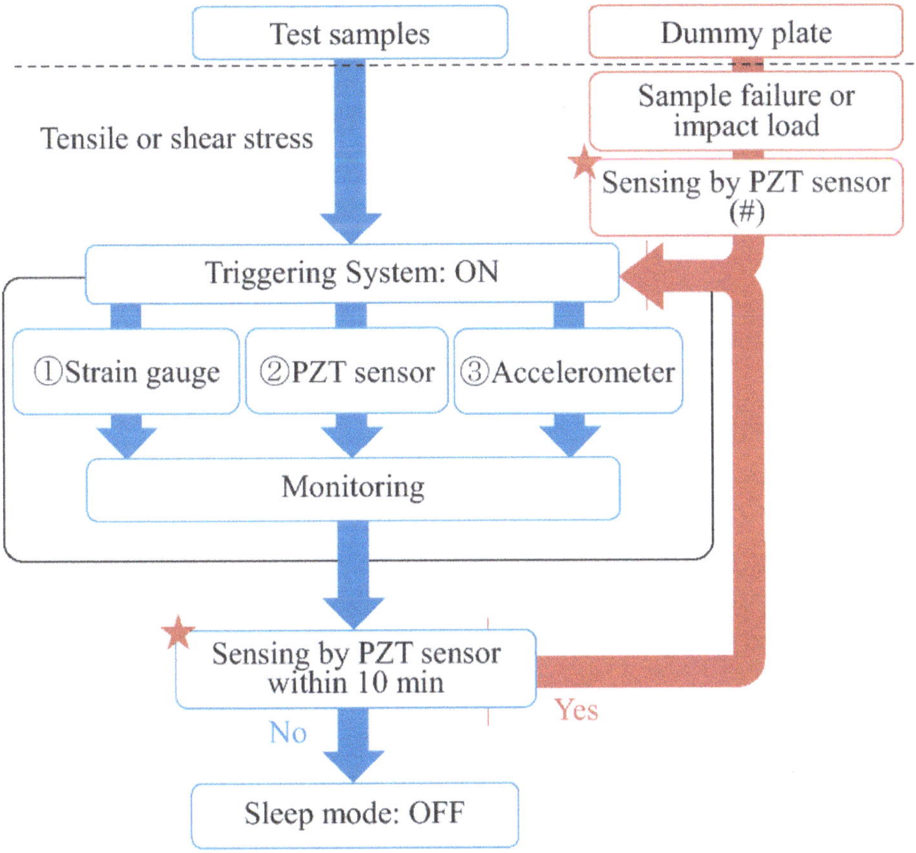

Fig. 2: Medical Data Transmission using IoT, Block

The private key x that used in the decryption process at the receiver side is encrypted using AES algorithm and sent to the receiver in an encrypted form to increase the security level. The encryption process can be mathematically modeled as given in the following equations below.

Simulation Environment

The implementation of our proposed model was carried out using the MATLAB R2015a software running on a personal computer with a 2.27 GHz Intel (R) Core (TM) I3 CPU, 8 GB RAM and Windows 7 as the operating system.

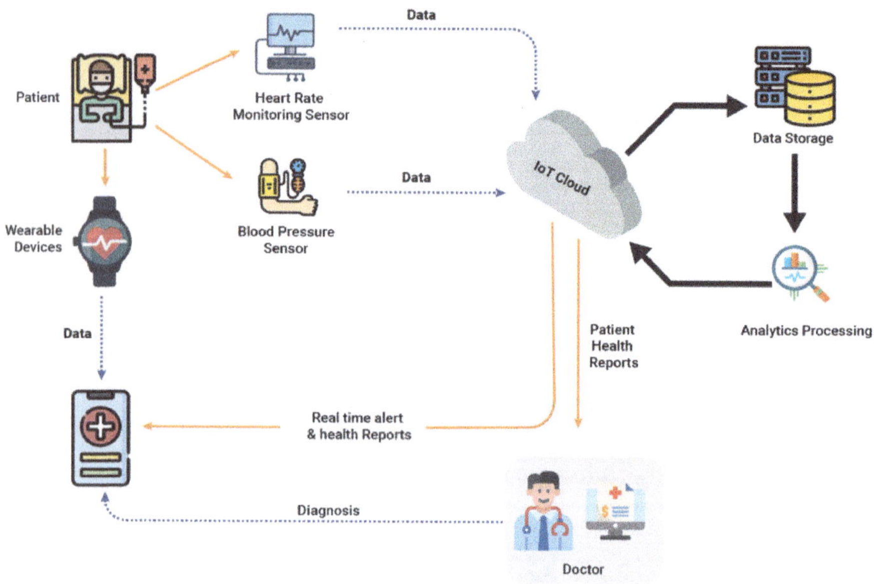

Fig. 3: Medical Data Transmission using IoT, Process

The calculation of specific statistical metrics determines the quality of the proposed security model. These metrics calculate the ratio between the original image and the stego image. The obtained results were evaluated based on six statistical parameters; the Peak Signal to Noise Ratio (PSNR), Mean Square Error (MSE), Bit Error Rate (BER), Structural Similarity (SSIM), Structural Content (SC), and Correlation.

Security Analysis

In this study, the comparisons were conducted between the original medical cover image and the stego-image. The proposed model was tested using different messages with different lengths and hiding them in both color and gry scale images.

The hidden message was analyzed before being transmitted and after being received by the expected recipient. That is to guarantee less distortion occurs within to the original cover file after concealing the secret text. Our model was applied to both datasets; DME eyes dataset [3][7] that represent the first three images, and DICOM dataset [3][8] that represent the last two images. the color and gray versions of these dataset.

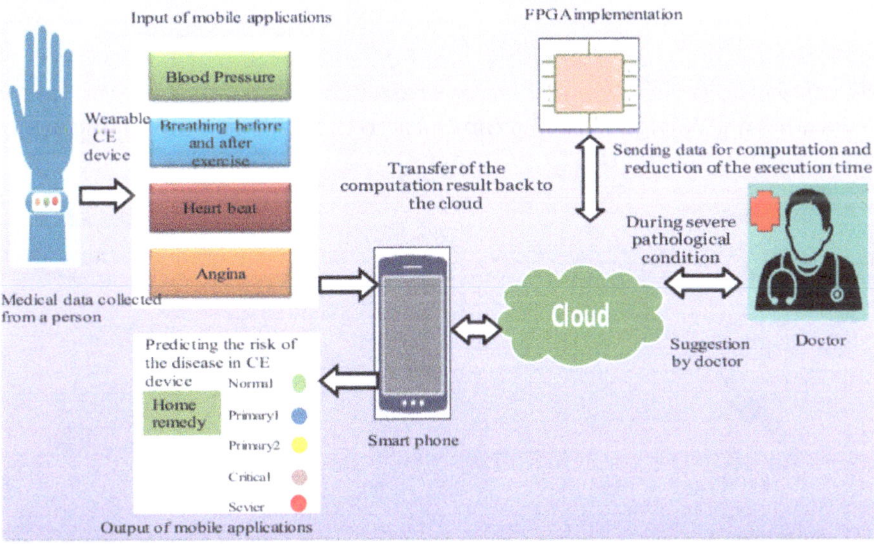

Fig. 4: Medical Data Transmission using IoT , Complete Structure/ Mobile Notifivation

In this model, the text is encrypted by using the evolved hybrid encryption scheme that was previously explained. Then it is being embedded using either 2D-DWT-1L or 2D-DWT2L stenography techniques. It was found that DWT-2L gives better PSNR and MSE results compared with DWT-1L in case of both color and gray scale images as illustrated in table I and table II respectively.

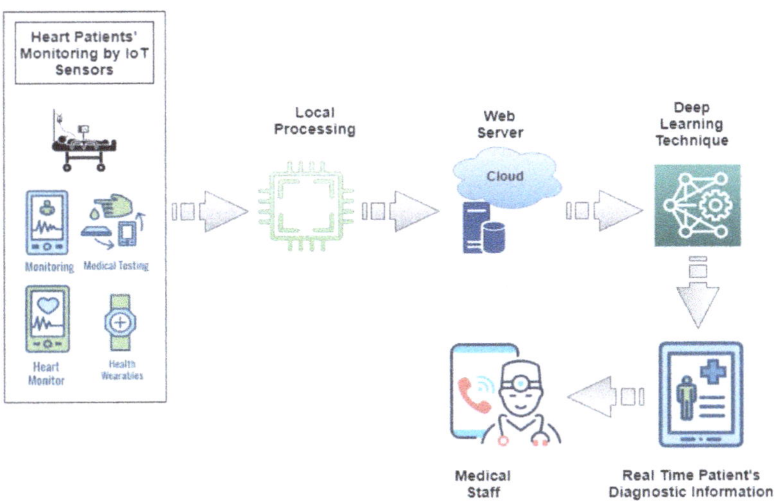

Fig. 5: Medical Data Transmission using IoT , Complete Structure/ Mobile Notifivation

The PSNR values were relatively reached to 57.44 and 56.39 in case of DWT-2L for color images and gray-scale images respectively. While the PSNR values were relatively reached to 56.13 and 55.43 in case of DWT-1L for color images and gray-scale images, respectively. The MSE values using DWT-2L varied from 0.12 to 0.57 for the color images and from 0.14 to 0.57 for the gray images.

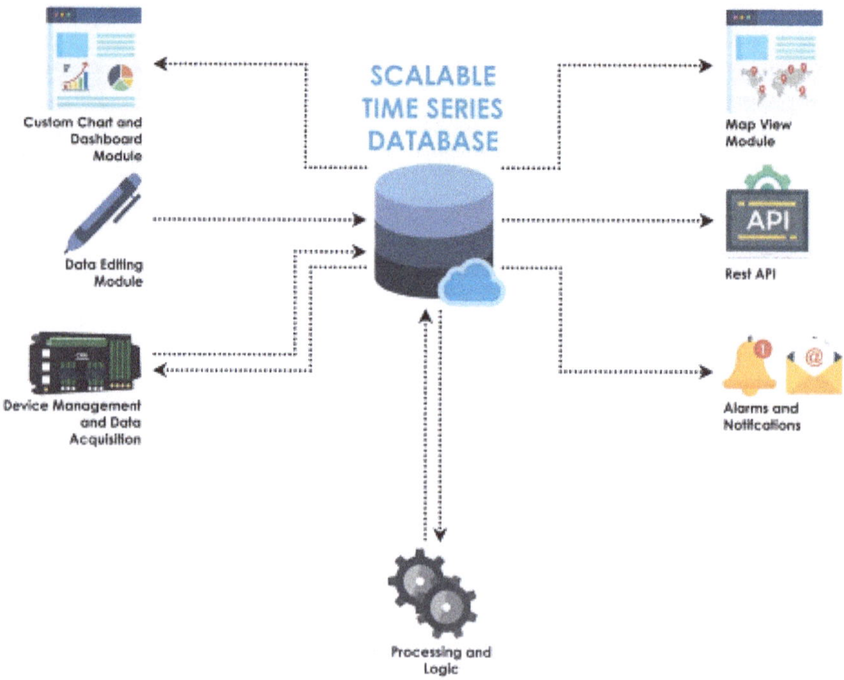

Fig. 6: Medical Data Transmission using IoT , Complete Structure

There was no variation between the studied images in the BER, where its values were zero for both images. The SSIM, SC, and Correlation were almost equal to one with all of the studied color and gray images as illustrated in table III and table IV respectively. Figures (6, 7, 8 and 9) show the histograms of the cover image before and after applying the proposed model on both color and gray scale images sequentially with different text sizes.

Comparing the Results

against another approach The performance of our model was compared with another technique developed by Anwar et al.[4] on 256x256 pixel medical color image using 18-byte text size. Table V shows the obtained PSNR and MSE values from applying our model as compared with those resulted by [4]. The results obtained after applying the models on 256*256 color medical images with text size 18 bytes. It was found that our proposed model had a higher PSNR value and a smaller MSE value that reveals the higher performance of our proposed model.

CONCLUSION

A secure patient's diagnostic data transmission model using both color and gray-scale images as a cover carrier for healthcare based IoT environment has been proposed. The proposed model engaged either 2D-DWT-1L or 2D-DWT-2L steganography and hybrid blending AES and RSA cryptographic techniques. The experimental results were evaluated on both color and gray-scale images with different text sizes. The performance was assessed based on the six statistical parameters (PSNR, MSE, BER, SSIM, SC, and correlation). Compared to the state-of-the-art methods, the proposed model proved its ability to hide the confidential patient's data into a transmitted cover image with high imperceptibility, capacity, and minimal deterioration in the received stego-image.

REFERENCES

[1] Ashraf Darwish, Aboul Ella Hassanien, Mohamed Elhoseny, Arun Kumar Sangaiah, Khan Muhammad, The Impact of the Hybrid Platform of Internet of Things and Cloud Computing on Healthcare Systems: Opportunities, Challenges, and Open Problems, Journal of Ambient Intelligence and Humanized Computing, 2022 (https://doi.org/10.1007/s12652- 017-0659-1)

[2] Abdulaziz Shehab, Mohamed Elhoseny, Khan Muhammad, Arun Kumar Sangaiah, Po Yang, Haojun Huang, Guolin Hou; Secure and Robust Fragile Watermarking Scheme for Medical Images, IEEE Access, Volume: PP, Issue: 99, (DOI: 10.1109/ACCESS.2018.2799240).

[3] Bairagi, A. K., Khondoker, R., & Islam, R. (2020). An efficient steganographic approach for protecting communication in the Internet of Things (IoT) critical infrastructures. Information Security Journal: A Global Perspective, 25(4-6), 197-212.

[4] Anwar, A. S., Ghany, K. K. A., & Mahdy, H. E. (2019). Improving the security of images transmission. International Journal, 3(4).

[5] Ahmed Abdelaziza, Mohamed Elhoseny, Ahmed S. Salama, A.M. Riad,"A Machine Learning Model for Improving Healthcare services on Cloud Computing Environment", Measurement, Volume 119, April 2018, Pages 117-128, 2019 (https://doi.org/10.1016/j.measurement.2018.01.022)

[6] Paschou, M., Sakkopoulos, E., Sourla, E., & Tsakalidis, A. (2020). Health Internet of Things: Metrics and methods for efficient data transfer. Simulation Modelling Practice and Theory, 34, 186-199.

[7] Muhammad Sajjad, Mansoor Nasir, Khan Muhammad, Siraj Khan, Zahoor Jan, Arun Kumar Sangaiah, Mohamed Elhoseny, Sung Wook Baik, "Raspberry Pi assisted face recognition framework for enhanced law-enforcement services in smart cities", Future Generation Computer Systems, Elsevier, 2020 (DOI: https://doi.org/10.1016/j.future.2017.11.013)

CITATIONS

- Ghai, D., Gianey, H.K., Jain, A. and Uppal, R.S., 2020. Quantum and dual-tree complex wavelet transform-based image watermarking. International Journal of Modern Physics B, 34(04), p.2050009.
- Jain, A. and Kumar, A., 2021. Desmogging of still smoggy images using a novel channel prior. Journal of Ambient Intelligence and Humanized Computing, 12(1), pp.1161-1177.
- Kumar, S., Jain, A., Shukla, A.P., Singh, S., Raja, R., Rani, S., Harshitha, G., AlZain, M.A. and Masud, M., 2021. A Comparative Analysis of Machine Learning Algorithms for Detection of Organic and Nonorganic Cotton Diseases. Mathematical Problems in Engineering, 2021.
- Misra, N.R., Kumar, S. and Jain, A., 2021, February. A Review on E-waste: Fostering the Need for Green Electronics. In 2021 International Conference on Computing, Communication, and Intelligent Systems (ICCCIS) (pp. 1032-1036). IEEE.
- Gupta, N., Jain, A., Vaisla, K.S., Kumar, A. and Kumar, R., 2021. Performance analysis of DSDV and OLSR wireless sensor network routing protocols using FPGA hardware and machine learning. Multimedia Tools and Applications, 80(14), pp.22301-22319.

Chapter - 03

Face Recognition Using IoT

Dr. Sanjay I. Nipanikar

Associate Professor, Flat. No.6, Virangula Residency, Dhabadi, Ambegaon (Bk). Pune. 411041.
sanjaynipanikar99@gmail.com

Dr. Makarand Shahade

Associate Professor, Department of Computer Engineering
SVKM's Institute of Technology, Dhule-424001

Dr. L Smitha

Dept. of IT, G. Narayanamma Institute of Technology and Science,
Shailkpet, Hyderabad, India.

ABSTRACT

Our Invention "Face Recognition and Identification using IoT and Deep Learning System" Our examination work involves the philosophy of fostering the containerised AI model. We have picked the methods for preparing the calculation to such an extent that it distinguishes the countenances caught by our camera, which is associated with the assistance of CSI connector. The calculation incorporates the idea of Deep Learning which is a subset of Artificial Intelligence. The strategy comprises of a few stages, for instance, Deep taking in Algorithm identifies the appearances from the picture, and afterward the picture is changed over to a bunch of angles. These angles can be changed over again to milestones to consider the central places of the picture and afterward the preparation step is performed utilizing the Support Vector Machine classifier. At last, the approved client is perceived. Our exploration work involves the strategy of fostering the containerised AI model and conveying the containerised application on the Raspberry Pi (IoT gadget), which comprises of the ARM processor. We infer that the containerised application run with high effectiveness, is versatile and adaptable between various stages, and the containerised application is viable with different structures (ARM, x86, amd64).

***Key:** Face, Recognition, IoT, ARM, x86, amd64, Deep Algorithm.*

FIELD OF THE INVENTION

Our Invention is related to a Face Recognition and Identification using IoT and Deep Learning System.

BACKGROUND

The basic contrast is that holders run on the host OS in client space, rather than an altogether unique climate, as VMs. Along these holders are lighter in weight, which makes them broadly more modest than a virtual machine. They can run relating to different applications in customer space, exist along with virtual conditions, and surprisingly run inside no less than one VMs.

Compartments are crucial to pushing information to the edge. Many edge devices are worked with process abilities to do extensively more than conventional advancement of data. They can examine approaching information streams utilizing prepared AI models.

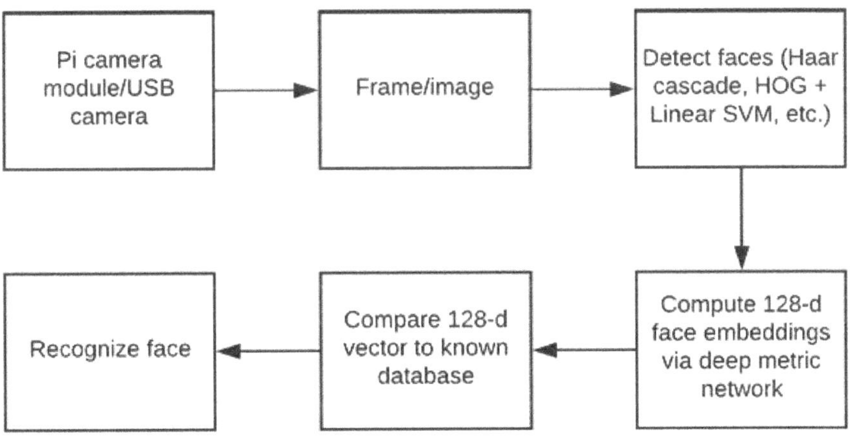

Fig. 1: Face Recognition and Identification using IoT and Deep Learning System, Flow Chart.

Thinking about AI applications, for instance, cameras and visual entryways can recognize circumstances rapidly and ready administrators as opposed to sending information to a focal area. In the current execution, we have fostered a containerized AI-based Face

Recognition model utilizing profound learning strategies. Profound Learning (DL) assumes a urgent part in Artificial Intelligence. Profound Learning is a subset of AI.

DL is testing, a popular exploration space of AI. About Facial Recognition, profound Learning engages us to achieve more unmistakable exactness than customary AI systems. With customary AI methods, hand-coding is needed for picture location and extraction, while this not required with profound Learning.

It is the AI that comprehends the necessities and addresses formative requirements while zeroing in on maintainability. Generally profound Learning utilizes the neural organization structures. Subsequently it very well may be known as profound neural organizations.

The primary objective of the Artificial Intelligence is to make wise machines that can think and settle the errands without the human direction. The security applications should have learning capacity which can learn dependent on the past experiences. Numerous occurrences can be run on a solitary part in a working framework utilizing the compartment based virtualization.

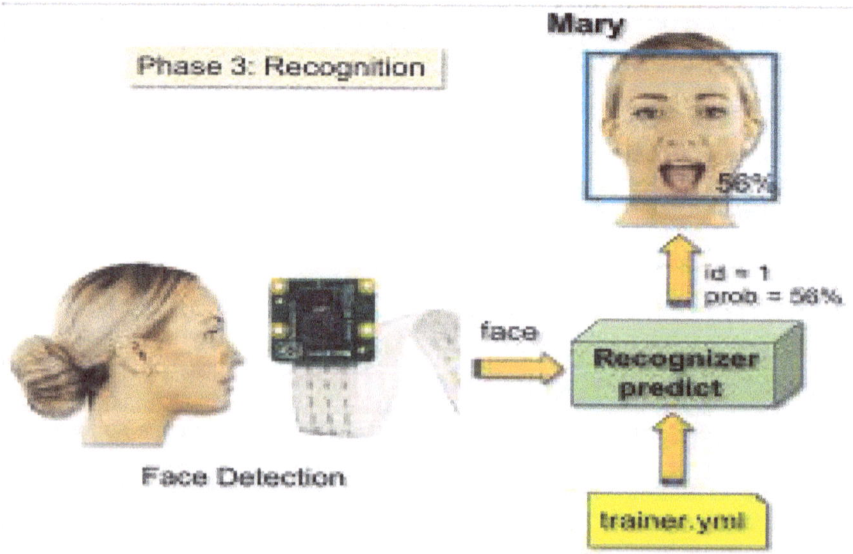

Fig. 2: Face Recognition and Identification using IoT and Deep Learning System.

The holder based virtualization further develops the application execution. Since every one of the applications can be run on a similar piece, there is asset productivity in this methodology, and it is not difficult to relocate.

The driving inspiration to begin this proposition is to examine the capacity and attainability to convey our containerised AI-put together face acknowledgment model with respect to Firefly-RK3399 and Raspberry Pi (IoT gadget).

The proposition is separated into two classifications. The First class is to plan the containerised AI based Face acknowledgment application for perceiving the approved client. There are various strategies in calculation advancement. The subsequent class incorporates checking the chance of fostering the model to such an extent that it is viable with numerous designs for instance, ARM (Firefly's engineering).

OBJECTIVES

The fundamental target of the proposal is to foster an AI-based face acknowledgment Model (which is carried out after the Deep Learning algorithm) for the security framework for settling on choices to lock or open the entryway framework and to convey the created AI Model in a Docker Container on an IoT stage.

The fundamental point of the proposition is accomplishing the edge registering idea that brings the Artificial Intelligence (through our AI model) to the low power Internet of Things (IoT) gadgets with the assistance of containerization idea. Containerization would be like the virtualizations.

Docker compartments are not difficult to port on different IoT gadgets (Firefly rk3399). Alongside the transportability, Docker incorporates every one of the conditions and modules needed for running the application in a compartment.

The objective of the invention is to provide a "Face Recognition and Identification using IoT and Deep Learning System" Our examination work involves the philosophy of fostering the containerised AI model.

The other objective of the invention is to provide a picked the methods for preparing the calculation to such an extent that it distinguishes the countenances caught by our camera, which is associated with the assistance of CSI connector.

The other objective of the invention is to provide a calculation incorporates the idea of Deep Learning which is a subset of Artificial Intelligence. The strategy comprises of a few stages, for instance, Deep taking in Algorithm identifies the appearances from the picture, and afterward the picture is changed over to a bunch of angles.

The other objective of the invention is to provide a angles can be changed over again to milestones to consider the central places of the picture and afterward the preparation step is performed utilizing the Support Vector Machine classifier. At last, the approved client is perceived.

The other objective of the invention is to provide a exploration work involves the strategy of fostering the containerised AI model and conveying the containerised application on the Raspberry Pi (IoT gadget), which comprises of the ARM processor.

The other objective of the invention is to provide a containerised application run with high effectiveness, is versatile and adaptable between various stages, and the containerised application is viable with different structures (ARM, x86, amd64).

SUMMARY

The principal objective is to ensure the machines begin to get information all alone with no help from the developer and they adjust to changing circumstances and given activities. Simulated intelligence and AI are generally utilized reciprocally, especially inside the domain of colossal data. In any case, this is certainly not a comparative variable, and it's important to get a handle on how these are frequently applied in any case.

Man-made consciousness could be a more extensive origination than AI that tends to the work of PCs to imitate the mental element elements of people. Right when machines take care of tasks subject to computations in a more intelligent manner, that is AI.

AI could be a great deal of Artificial Intelligence and spotlights on the force of machines to get a social occasion of information and find out on their own, routinely changing estimations as they come out as comfortable with the information, they are taking care of.

Like ML, "profound Learning" is in like manner a method that concentrates features or attributes from rough informational collections. The focal matter of differentiation is DL does this by utilizing a multi-facet Artificial neural organization with many covered layers stacked in a consistent movement. DL moreover has, somewhat, logically present day calculations and requires all the more impressive computational assets.

These are incredibly made PCs with elite execution CPUs or GPUs. Most profound learning methods use neural organization plans, which is the explanation profound learning models are routinely suggested as profound neural frameworks. The articulation "profound" signifies the amount of covered layers in the neural organization.

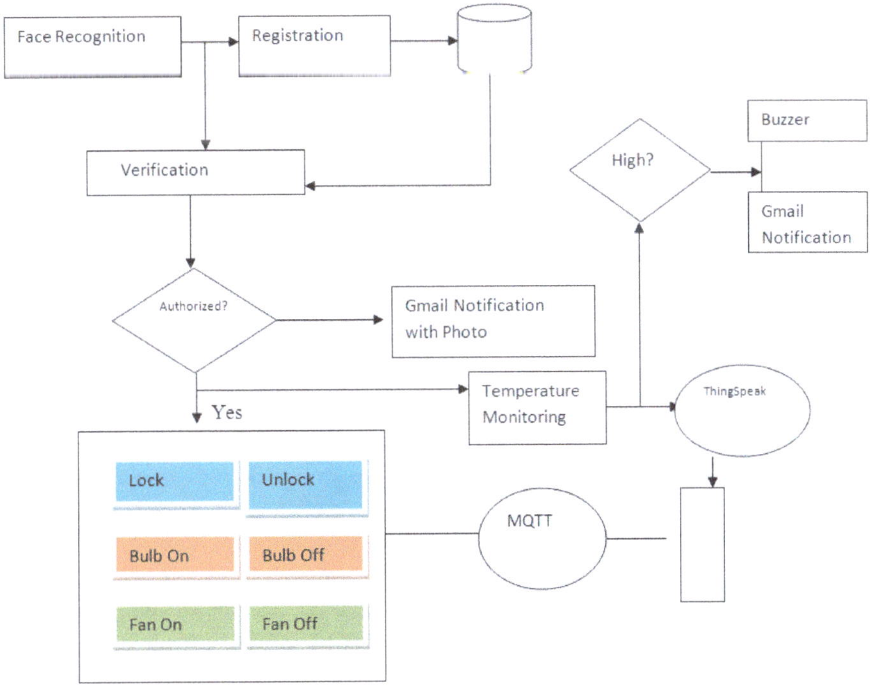

Fig. 3: Face Recognition and Identification using IoT and Deep Learning System,

Ordinary neural frameworks simply hold back 2-3 secret layers, while profound neural organizations can have as much as 150. Artificial neural organizations (ANNs) are processing frameworks that are really enlivened by natural neural organizations.

Artificial neural organizations (ANNs) are the frameworks inside the neural organizations. Such systems learn (legitimately upgrade their ability) to tackle tasks by pondering points of reference, overall without undertaking express programming. Normally, neurons are coordinated in layers.

Later some time, thought focused on organizing unequivocal intellectual abilities, inciting deviations from science, for instance, back proliferation, or passing information in the change bearing and adjusting the framework to reflect that information.

DIAGRAM

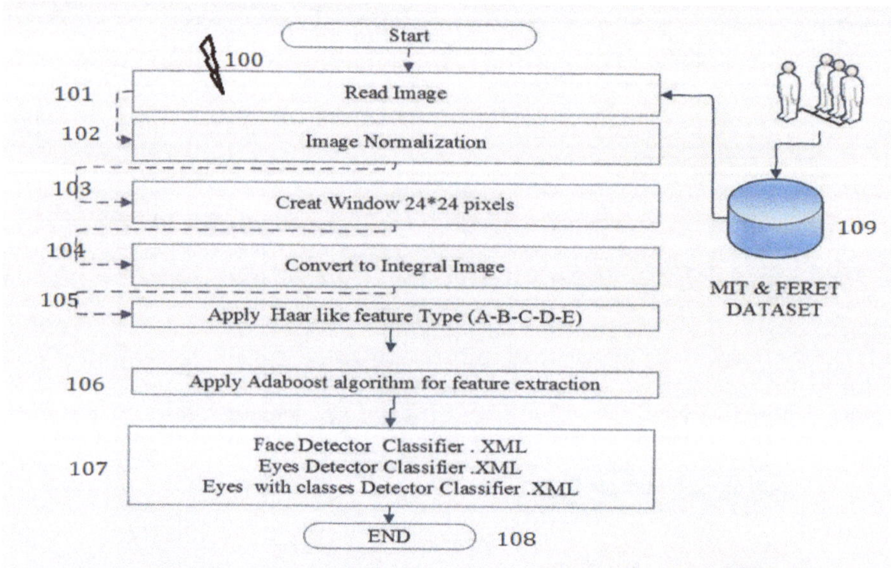

Fig. 4: Face Recognition and Identification using IoT and Deep Learning System, Flow Chart.

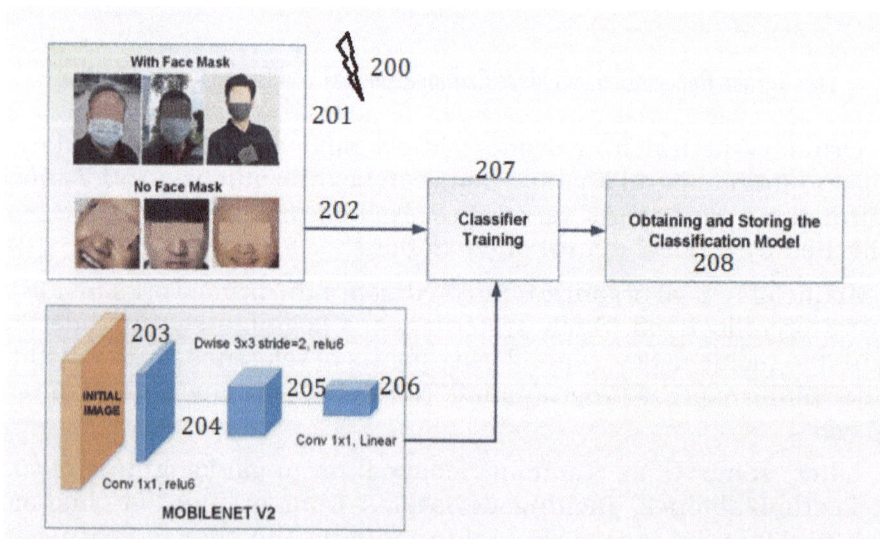

Fig. 5: Face Recognition and Identification using IoT and Deep Learning System, Block Diagram.

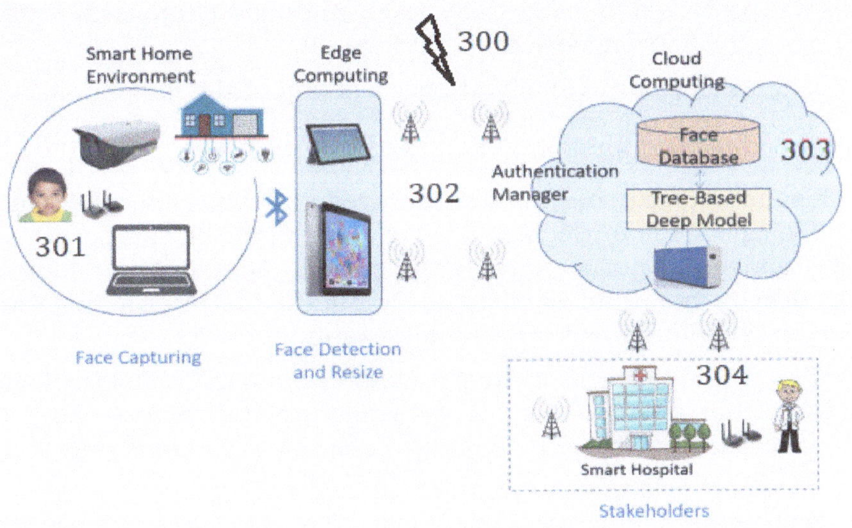

Fig. 6: Face Recognition and Identification using IoT and Deep Learning System

DESCRIPTION

It manages constant acknowledgment of appearances by request of pictures being recorded in a camcorder. The main utilization of face goal is chiefly for managing the security prerequisite. This part involves distinguishing, recording, and diagnosing of appearances.

Bountiful endeavors were advanced on face acknowledgment for a 2-Dimensional (2D) power pictures. However, a total report hasn't been done in recognizing an undividable by achieving different strategies for detecting like 3-Dimensional (3D) or reach information IR symbolism.

The above figure addresses the framework level engineering where the gadget utilized for the execution of the undertaking is Raspberry Pi 3 model B+, The working framework utilized is raspberries and the working framework generally separates into client namespace and bit namespace, the Docker motor is introduced on the working framework with upheld ARM design pairs to run the Docker motor to have holders.

Libraries and modules and conditions that are needed to run the AI model face acknowledgment application on IoT stage are packaged inside the compartment to give the seclusion from different conditions. Application concludes the credibility of the individual's face contingent upon the perceived face certainty level, the higher the certainty level the higher the genuineness of the individual.

For the most part, IoT cloud-based is one of the methodologies in IoT world, where the calculation of the application and information is being broke down and checked on the cloud.

This methodology has both advantaged and hindered, where in our methodology one of our central goal is to bring the calculation close to the IoT gadget, in not many case the calculation should be done over IoT gadget rather than sending every one of the information to the cloud for calculation and information examination.

IoT cloud-based methodology is Homogenous methodologies where the stage that is handling and investigating the information has a normalized equipment with generally explicit CPU engineering stages.

This model of approach has homogenous equipment, comparative standard correspondence conventions, the application layer might vary for one framework to other framework and administration the executives are normalized, this makes simple for general applications where information from the IoT gadgets are spilled to cloud motors for calculation, information investigation, process the information and to be checked.

In this methodology IoT gadgets simply stream the gathered information from sensors and other fringe gadgets and convey over standard IoT conventions like MQTT, CoAP and so on to send the information to the cloud motors.

The principle benefits and disservices are recorded beneath Advantages –

Checking Services - Highly Accessible Cloud stage - Device network stage - Fast information handling and investigation.

The Artificial Intelligence face acknowledgment model is intended to identify, catch, and perceive the Face from a picture.

The model is carried out to such an extent that it works (runs) on various processors like ARM, AMD, X86, Intel, and so on The principle objective is to foster face acknowledgment AI Model that is appropriate to settle on choices for locking or opening the entryway framework as a utilization case and to send the created AI Model in a Docker Container.

This model can settle on choices dependent on preparing and dataset. Numerous countenances and names are planned and are utilized as preparing sets to the framework.

This acknowledgment model framework utilizes profound learning and picture handling strategies to distinguish the Face in the live transfer video and interaction the edges with the matching countenances from the datasets. In such cases advancements like AI and preparing datasets are utilized to train the framework to do the means in process.

The cycle incorporates preparing diverse AI calculations are taken and anchored together to get best outcomes. The initial step is to recognize the appearances in the edge or an image, this is an extraordinary component for the framework where the framework can consequently choose faces disregarding different foundations and colors and furthermore it can ensure that the distinguished Face is in great concentration before it recognizes the Face.

To begin observing countenances in a casing, the casing is changed over in to highly contrasting picture on the grounds that to diminish the size of the casing that aides in expanding the handling speed then each and every pixel in the edge at single time is viewed as that are straightforwardly encompassing it.

RESULT

Our Invention "Face Recognition and Identification using IoT and Deep Learning System" Our examination work involves the philosophy of fostering the containerised AI model. We have picked the methods for preparing the calculation to such an extent that it distinguishes the countenances caught by our camera, which is associated with the assistance of CSI connector. The calculation incorporates the idea of Deep Learning which is a subset of Artificial Intelligence. The strategy comprises of a few stages, for instance, Deep taking in Algorithm identifies the appearances from the picture, and afterward the picture is changed over to a bunch of angles. These angles can be changed over again to milestones to consider the central places of the picture and afterward the preparation step is performed utilizing the Support Vector Machine classifier. At last, the approved client is perceived. Our exploration work involves the strategy of fostering the containerised AI model and conveying the containerised application on the Raspberry Pi (IoT gadget), which comprises of the ARM processor. We infer that the containerised application run with high effectiveness, is versatile and adaptable between various stages, and the containerised application is viable with different structures (ARM, x86, amd64).

- According to claim1# the invention is to a "Face Recognition and Identification using IoT and Deep Learning System" Our examination work involves the philosophy of fostering the containerised AI model.

- According to claim1,2,3# the invention is to a picked the methods for preparing the calculation to such an extent that it distinguishes the countenances caught by our camera, which is associated with the assistance of CSI connector.
- According to claim1,2,3# the invention is to a calculation incorporates the idea of Deep Learning which is a subset of Artificial Intelligence. The strategy comprises of a few stages, for instance, Deep taking in Algorithm identifies the appearances from the picture, and afterward the picture is changed over to a bunch of angles.
- According to claim1,2,3,4# the invention is to a angles can be changed over again to milestones to consider the central places of the picture and afterward the preparation step is performed utilizing the Support Vector Machine classifier. At last, the approved client is perceived.
- According to claim1,2,3,4# the invention is to a exploration work involves the strategy of fostering the containerised AI model and conveying the containerised application on the Raspberry Pi (IoT gadget), which comprises of the ARM processor.
- According to claim1,2,3,4# the invention is to a containerised application run with high effectiveness, is versatile and adaptable between various stages, and the containerised application is viable with different structures (ARM, x86, amd64).

REFERENCES

[1] Y. Januzaj, A. Luma, Y. Januzaj, V. Ramaj., "Real time access control based on face recognition," in International Conference on Network security & Computer Science (ICNSCS-15), pp. 7-12, 2022.

[2] M. Sahani, C. Nanda, A. K. Sahu, B. Pattnaik., "Home security system based on face recognition,"" 2020 Int. Conf. Circuits, Power Comput. Technol. [IC-CPCT-2015], pp. 1–6, 2015.

[3] G. Senthilkumar, K. Gopalakrishnan, V. S. Kumar., "Embedded image capturing system using raspberry pi system," vol. 3, No. 2, pp. 213-215, 2019.

[4] M. R. Mulla., "Facial image based security system using PCA," pp. 548-553, 2015.

[5] M. H. Jusoh & F. Bin Jamali, "Home security system using internet of things," 2017. [6] S. S. Liew, M. Khalil-Hani, S. Ahmad Radzi, R. Bakhteri., "Gender classification: A convolutional neural network approach," Turkish J. Electr. Eng. Comput. Sci., vol. 24, No. 3, pp. 1248-1264, 2016.

[6] M. Sajjad et al., "Raspberry pi assisted face recognition framework for enhanced law-enforcement services in smart cities," Futur. Gener. Comput. Syst., 2019.

[7] A. R. Syafeeza, S. S. Liew, R. Bakhteri., "Convolutional neural networks with fused layers applied to face recognition," Int. J. Comput. Intell. Appl., vol. 14, No. 3, 2015.

[8] A. R. Syafeeza, M. Khalil-Hani, S. S. Liew, R. Bakhteri., "Convolutional neural network for face recognition with pose and illumination variation," Int. J. Eng. Technol., Vol. 6, No. 1, pp. 44-57, 2014.

[9] K. Syazana-Itqan, A. R. Syafeeza, N. M. Saad, N. A. Hamid, W. H. Bin Mohd Saad., "A review of finger-vein biometrics identification approaches," Indian J. Sci. Technol., vol. 9, No. 32, 2019.

[10] S. Ahmad Radzi, M. Khalil-Hani, R. Bakhteri., "Finger-vein biometric identification using convolutional neural network," Turkish J. Electr. Eng. Comput. Sci., vol. 24, No. 3, pp. 1863-1878, 2016.

[11] S. Ahmad Radzi., "A MATLAB-based convolutional neural network approach for face recognition system," J. Bioinformatics. Proteomics Rev, vol. 2(1), pp. 1-5, 2016.

CITATIONS

- Jain, J., Jain, A., Srivastava, S. K., Verma, C., Raboaca, M. S., & Illés, Z. (2022). Improved Security of E-Healthcare Images Using Hybridized Robust Zero-Watermarking and Hyper-Chaotic System along with RSA. Mathematics, 10(7), 1071.
- Jain, J., & Jain, A. (2022). Securing E-Healthcare Images Using an Efficient Image Encryption Model. Scientific Programming, 2022.
- Kumar, S., Shailu, A., Jain, A., & Moparthi, N. R. (2022). Enhanced Method of Object Tracing Using Extended Kalman Filter via Binary Search Algorithm. Journal of Information Technology Management, 14(Security and Resource Management challenges for Internet of Things), 180-199.
- Jain, A., Dwivedi, R. K., Alshazly, H., Kumar, A., Bourouis, S., & Kaur, M. (2022). Design and Simulation of Ring Network-on-Chip for Different Configured Nodes. CMC-COMPUTERS MATERIALS & CONTINUA, 71(2), 4085-4100.
- Sharma, S.K., Jain, A., Gupta, K., Prasad, D. and Singh, V., 2019. An internal schematic view and simulation of major diagonal mesh network-on-chip. Journal of Computational and Theoretical Nanoscience, 16(10), pp.4412-4417.

CHAPTER - 04

Predictors of Embolic Stroke Patients Using IoT

Neeraj Milind Shahane

CEO, D-Tech, Philter Co, U1/26 Arnott Street Clayton VIC 3168
Email : shahaneneeraj00@gmail.com

V. M. Subramanyam

Asst. Prof., Vijaya Engineering College

Prof. (Dr.) Vivek Kumar

Vice Chancellor, Quantum University, Roorkee, UK, India.

Prof (Dr.) B.K. Sarkar

MES's, Pillai HOC College of Engineering & Technology,
Rasayani Taluka Panvel, Dist, Navi Mumbai, Maharashtra 410207, India.

ABSTRACT

Our Books Chapter "Predictors of Embolic Stroke Patients using IoT" is a review cross-sectional investigation of patients determined to have cryptogenic stroke at an exhaustive stroke place and alluded for something like 21 days of drawn out short term telemetry. Telemetry reports and information from the patient's stroke hospitalization, including imaging studies, electrocardiogram (EKG), echocardiogram, important bodily functions, and research facility information, were surveyed. A modest amount of the 121 patients remembered for the review were determined to have AF dependent on short term telemetry. There was a solid relationship between presence of untimely atrial compressions (PACs) on confirmation EKG and resulting recognition of AF (P = .004). Enormous left *atrial distance across on echocardiogram was corresponded with AF recognition in guys (P = 0.024). Nonetheless, there was no relationship among AF and other echocardiographic estimations. Thyroid-animating chemical (TSH) levels were essentially higher in patients with cryptogenic stroke having AF (P = .008), with a TSH more noteworthy than 4.20 mIU/L prescient of identification of AF (P < .001). RE-SPECT ESUS was a randomized, controlled preliminary (564 destinations, 42 nations) evaluating dabigatran versus ibuprofen for the avoidance of repetitive stroke in patients with embolic stroke of dubious source. Of 5390 patients enlisted and followed for a middle of 19 months, 403 (7.5%) were found to foster AF revealed as an unfriendly occasion or utilizing heart checking per standard clinical consideration. Univariable and multivariable relapse examinations were performed to characterize indicators of AF.*

Key: *Predictors, Embolic, Stroke ,Patients, IoT.*

FIELD

Our Invention is related to a Independent Predictors of Atrial Fibrillation in Cryptogenic Embolic Stroke Patients using IoT-Based System.

BACKGROUND

Numerous patients determined to have cryptogenic stroke or transient ischemic assault are in this manner found to have atrial fibrillation (AF) on short term cardiovascular telemetry checking. Distinguishing proof of prescient variables for the discovery of AF could help with patient determination to expand the yield of telemetry and hurry inception of fitting auxiliary stroke avoidance.

An extent of patients with embolic stroke of unsure source have quiet atrial fibrillation (AF) or foster AF later the underlying assessment. Better comprehension of the danger for improvement of AF is basic to carry out ideal observing techniques determined to forestall intermittent stroke inferable from hidden AF.

The RE-SPECT ESUS preliminary (Randomized, Double-Blind Evaluation in Secondary Stroke Prevention Comparing the Efficacy and Safety of the Oral Thrombin Inhibitor Dabigatran Etiolate Versus Acetylsalicylic Acid in Patients with Embolic Stroke of Undetermined Source) gives a chance to survey indicators for creating AF and related repetitive stroke.

Atrial fibrillation (AF) is a strange cardiovascular beat portrayed by a complicated atrial movement. AF is perceived in the electrocardiogram (ECG) as a sporadically unpredictable beat enduring more than

30 s, with no noticeable P-waves going before the QRS complex. AF pervasiveness is age, sexual orientation, and race subordinate. It is especially high in the old populace, arriving at 10–17% in subjects 80 years and more seasoned.

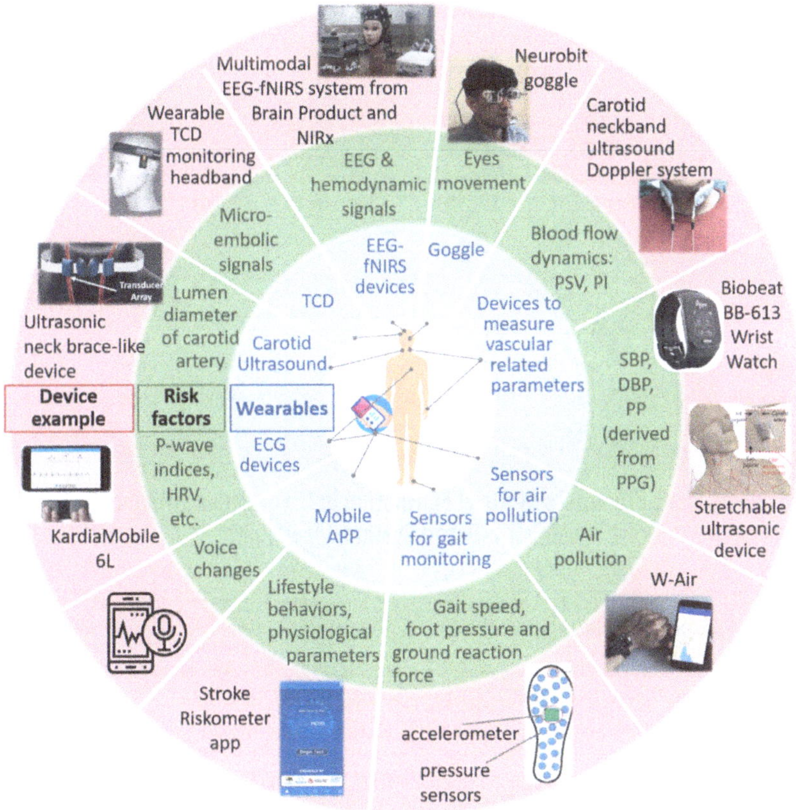

Fig. 1: Independent Predictors of Atrial Fibrillation in Cryptogenic Embolic Stroke Patients using IoT-Based System, Flow Chart.

Likewise, AF is more pervasive in guys and in the white population.3 AF is related with critical horribleness and mortality. One out of five strokes is related with AF and 33% of cardiovascular arrhythmias hospitalizations are because of AF-related complexities. AF has been related with a twofold expansion in the danger of death. Also, the maturing populace in the US and overall is prompting a uniquely expanding AF commonness.

The high pervasiveness of asymptomatic AF has huge clinical ramifications on the finding and the executives of AF. Discontinuous

ECG assessment during clinical visits has a low probability of recognizing paroxysmal AF. Constant checking would expand the odds of AF location, subsequently permitting fitting essential and optional stroke anticipation methodologies to diminish the high grimness and mortality of stroke.

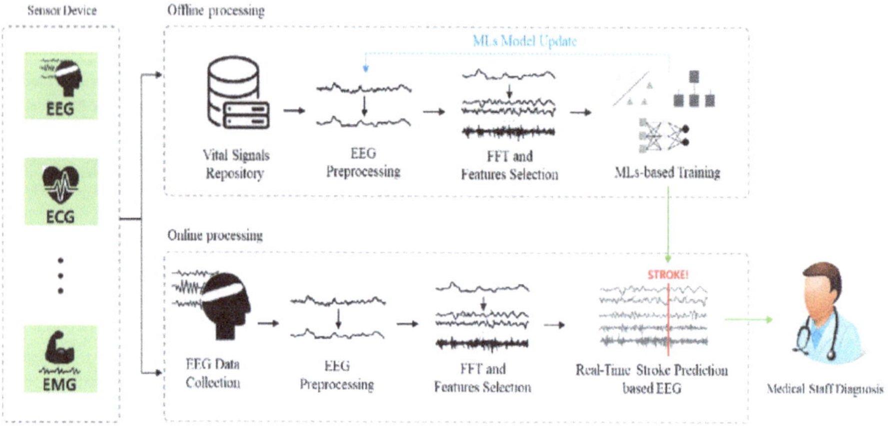

Fig. 2: Independent Predictors of Atrial Fibrillation in Cryptogenic Embolic Stroke Patients using IoT-Based System, Flow Chart.

For patients with intense ischemic stroke or transient ischemic assault, roughly 10% will have new AF distinguished during their clinic affirmation. Constant ECG checking for 30 days is suggested if there should be an occurrence of an embolic stroke of dubious reason (cryptogenic).

Novel non-meddlesome methodologies for heart cadence observing might possibly empower early and exact recognition of asymptomatic paroxysmal AF and make a change in AF the board. Particularly for asymptomatic AF cases, new instruments that permit the AF location will help make the.

This outlines the utilization and capability of IoT, especially in the medical care area. These days, everybody in the public arena is so distracted with adapting to their life that their wellbeing is being dismissed. With the presentation of savvy sensors, it is currently attainable to persistently screen a singular's activities and save information just as expect the event of a coronary episode before the patient encounters the impacts of a cardiovascular failure. Accordingly, it is basic to choose and utilize the right sensors.

OBJECTIVES

The objective of the invention is to provide a "Independent Predictors of Atrial Fibrillation in Cryptogenic Embolic Stroke Patients using IoT-Based System" This was a review cross-sectional investigation of patients determined to have cryptogenic stroke at an exhaustive stroke place and alluded for something like 21 days of drawn out short term telemetry.

The other objective of the invention is to provide a reports and information from the patient's stroke hospitalization, including imaging studies, electrocardiogram (EKG), echocardiogram, important bodily functions, and research facility information, were surveyed.

The other objective of the invention is to provide a modest amount of the 121 patients remembered for the review were determined to have AF dependent on short term telemetry. There was a solid relationship between presence of untimely atrial compressions (PACs) on confirmation EKG and resulting recognition of AF (P = .004). Enormous left atrial distance across on echocardiogram was corresponded with AF recognition in guys (P = .024).

The other objective of the invention is to provide a Nonetheless, there was no relationship among AF and other echocardiographic estimations. Thyroid-animating chemical (TSH) levels were essentially higher in patients with cryptogenic stroke having AF (P = .008), with a TSH more noteworthy than 4.20 mIU/L prescient of identification of AF (P < .001). RE-SPECT ESUS was a randomized, controlled preliminary (564 destinations, 42 nations) evaluating dabigatran.

The other objective of the invention is to provide a versus ibuprofen for the avoidance of repetitive stroke in patients with embolic stroke of dubious source. Of 5390 patients enlisted and followed for a middle of 19 months, 403 (7.5%) were found to foster AF revealed as an unfriendly occasion or utilizing heart checking per standard clinical consideration. Univariable and multivariable relapse examinations were performed to characterize indicators of AF.

SUMMARY

The Institutional Committee and a banded together emergency clinic endorsed the information assortment design and trial strategies to tentatively assemble nitty gritty data on intense stroke patients to lead this review. A marked informed assent structure was expected of every tolerant. The total information assortment technique was completed as per the endorsed system and guidelines.

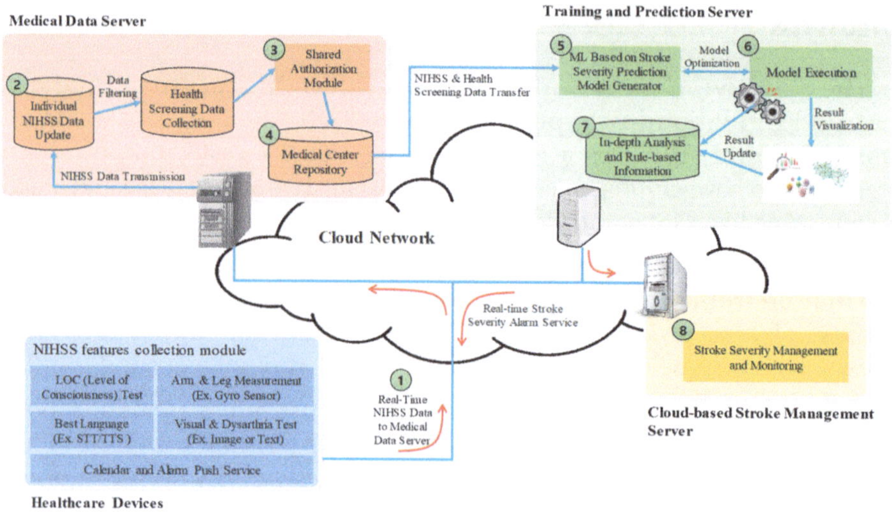

Fig. 3: Independent Predictors of Atrial Fibrillation in Cryptogenic Embolic Stroke Patients using IoT-Based System, Flow Chart.

As this is a review study, the consideration and avoidance measures are outlined remembering the goal of the review, which are as per the following.

- **Consideration models:** Those patients that were presently selected a recovery plan, just as the people who were neuropsychological and clinical examinations, were just included. Second, people with intense stroke, who were conceded in emergency clinic inside 24 hours from January 2021 to June 2021, were likewise concentrated tentatively (n = 90).
- **Rejection**: Patients with suggestive cardiovascular breakdown, failure to gather pulse signals inside 48 hours later thought, or pulse signs of exceptionally low quality or relics were excluded from the review.

Data Acquisition

For clear review, information relating to age, sex, instruction, and conjugal status of volunteers were gathered. Alongside it, the Barthes file was determined for every one of the patients.

It is an ordinal scale that is most generally used to evaluate the practical freedom in the spaces of individual consideration and versatility in patients with ongoing, handicapping conditions, particularly in the recovery settings.

It is generally used to decide the seriousness of a stroke and intellectual capacity. The composite BI is a 100-point scale, with 100 showing total autonomy, 60–99 recommending little dependence, 40–59 demonstrating moderate dependence, 20–39 showing extreme dependence, and 0–19 showing absolute dependence. The useful autonomy of the patients was evaluated utilizing the action.

It was guaranteed that all patients got normal basic consideration, which included pulse, pulse, and EEG observing. Here, the glove estimates the pulse and circulatory strain, though EEG was estimated with a convenient EEG machine. Later on, we intend to set up a savvy glove that does constant EEG.

Heart Rate and EEG

Pulse and EEG sensors were utilized to plan the contraption. Arduino Uno was joined for control activity, while the GSM module investigated correspondence control. The pulse and EEG sensors ideal sense people's pulse. The square graph of our glove-based pulse and EEG observing framework. The microcontroller contrasted the detected vitals and the edge esteem consistently. The provided anodes were separated from one another by 10% or 20% of the reference distance, as per the typical strategy.

At the point when the detected vitals change, the microcontroller perceives the change and sends a programmed barcoded ID message to the patient's primary care physician or family through the standard GSM module associated with the regulator unit (this component is as of now under some minor changes). The other advantage of it is there is an arrangement where the specialists can undoubtedly get to the refreshed patient's record from the data set, along these lines choosing further mediation.

Obviously, the pulse sensor detects our heartbeat and converts it into electrical information, which are obtained from Arduino associated with it, and this Arduino sends the handled information to another Arduino utilizing the Wi-Fi module that is associated with home apparatuses. Here, the second Arduino processes the information dependent on the information got from the primary Arduino, and the second Arduino will be modified to such an extent that it naturally controls the home machines dependent on the handled information.

In the transmitter part of the circuit, the accelerometer sensor detects the development of hand and records the signs and converts the signs into electrical information that are then handled and broke down by the Arduino found at the beneficiary end, and this information are communicated to the collector circuit. The pulse sensor and accelerometer are executed close by gloves or a watch, which is wearable by any individual.

Information portrayal was completed by a UI where on demand the information is first recovered from the cloud data set, and if necessary it is additionally handled, and afterward it is shown for additional either investigation or understanding. To get criticism of framework disappointment, the glove was likewise outfitted with LEDs (securing or transmission).

Atrial fibrillation (AF) is a heartbeat issue related with expanded dreariness and mortality. It is the main danger factor for cardio embolic stroke and its initial identification is vital in both essential and optional stroke counteraction. Constant checking of heart musicality is today conceivable because of shopper grade wearable gadgets, empowering ground-breaking analytic and patient administration devices. Such checking is conceivable utilizing minimal expense simple to-carry out optical sensors that today prepare most of wearables.

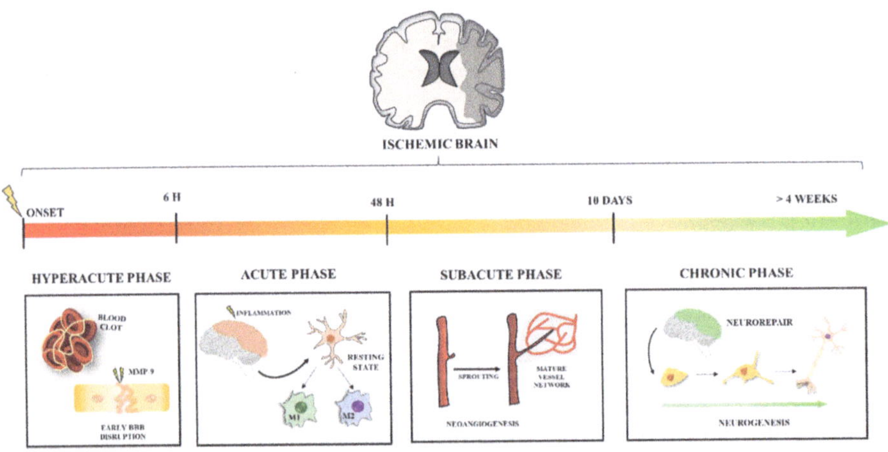

Fig. 4: Independent Predictors of Atrial Fibrillation in Cryptogenic Embolic Stroke Patients using IoT-Based System, Flow Chart.

These sensors record blood volume varieties—an innovation known as photograph plethysmography (PPG)— from which the pulse and other physiological boundaries can be removed to illuminate about client action, wellness, rest, and wellbeing. As of late, new wearable gadgets were presented as being equipped for AF location, confirmed by huge imminent preliminaries sometimes.

Such gadgets would take into consideration early screening of AF and inception of treatment to forestall stroke. This audit is an outline of an assortment of work on AF location utilizing PPG. An intensive record of the sign handling, AI, and profound learning approaches utilized in these investigations is introduced, trailed by a conversation of their restrictions and difficulties towards clinical applications.

PPG waveform is produced during a heart cycle and regularly estimated at a fringe site. Thusly, it is basically a heartbeat pressure waveform that begins from the heart withdrawal and proliferates through the vascular tree. As blood stream is constrained by neural, cardiovascular, and respiratory associations, different physiological boundaries could hypothetically be removed from examining a PPG signal.

PPG waveforms have ordinary morphological parts relating to milestone occasions in the heart cycle. During the withdrawal of the left ventricle, blood is catapulted out of the heart and engenders along the blood vessel tree, this relates to the underlying positive slant of a PPG beat. The systolic pinnacle denotes the limit of the waveform.

A diminishing in adequacy following the systolic pinnacle is set apart by a neighbourhood least, or the dicrotic indent, which compares to the end of aortic valves isolating the systolic and diastolic stages. At times, a third pinnacle following the dicrotic indent can be distinguished. It relates to a reflected part of the forward wave from different reflection locales including vessel bifurcations.

Clinical Boundaries

One essential clinical utilization of PPG is blood vessel blood oxygen immersion (SpO2) assessment through beat oximetry. SpO2 is characterized as the level of oxygen immersion in the blood vessel blood, which can be estimated by the proportion of oxygenated hemoglobin fixation to the absolute hemoglobin focus, with an ordinary reach somewhere in the range of 97% and 98%.

As of late, new utilizations of PPG have arisen for the persistent assessment of important cardiovascular boundaries in wandering settings. Pulse, circulatory strain, and respiratory rate could be firmly observed for wellness or wellbeing appraisal. Progressed symptomatic uses of PPG were likewise visualized. Cardiovascular capacity, blood vessel solidness, autonomic sensory system (ANS) reactions, and apnea are among conditions that might actually be identified or assessed utilizing PPG.

Changes in blood volume are coordinated with the heart beats, such synchrony is showed by the concordance of between thump stretches (RR spans) estimated in PPG and time-synchronized ECG. Pulse fluctuation (HRV) is a roundabout estimation of ANS, and it has additionally been considered as a substitute boundary of the collaboration between the mind and cardiovascular framework.

HRV measurements can be gotten from examining RR stretches on schedule and additionally recurrence space just as utilizing nonlinear powerful examination draws near. Respiratory rate is one of the basic indispensable signs and still up in the air from the time–recurrence portrayal of a PPG signal.

A PPG signal has two principle parts: a semi static direct current (DC) part, which addresses light reflected/sent from static blood vessel blood, venous blood, skin and tissues; and pulsatile substitute current (AC) part which emerges from adjustment in light ingestion because of changes in blood vessel blood volume.

PPG estimation can be done utilizing two modes: transmission and reflectance. In transmission mode, the light communicated through the medium is recognized by a photodetector (PD), which is situated in the contrary site of the light source. The sensor should be situated on the body at a site where sent light can be distinguished. The estimation site is restricted to the limits of the body, like the fingertip or ear cartilage.

Lately, neurological sicknesses have turned into a champion among the wide range of various illnesses and are the main explanations behind mortality and horribleness from one side of the planet to the other. The current review's point is to direct a pilot review for testing the model of the planned glove-wearable innovation that could recognize and break down the pulse and EEG for better administration and keeping away from stroke outcomes.

The subjective, clinical test strategy for evaluation was investigated by joining utilization of an IoT-based constant surveying clinical glove that was planned utilizing pulse based and EEG-based sensors. We directed organized meetings with 90 patients, and the aftereffects of the meetings were investigated by utilizing the Barthel file and were assembled in like manner.

By and large, the extent of patients who followed appropriate day by day pulse recording conduct went from 46.9% in the main month of the preliminary to 78.2% following 3–10 months of the intercessions. In the meantime, the level of people having a sporadic pulse tumbled from 19.5% in the primary month of the preliminary to 9.1% following 3–10 months of intercession research.

In T5, we observed that delta relative power diminished by 12.1% and 5.8% contrasted and pattern at 3 and at 6 months and a normal increment was 24.3±0.08. Beta-1 remained moderately consistent, while theta relative power developed by 7% and alpha relative power expanded by 31%. The T1 side of the equator had more prominent mean upsides of delta and theta relative power than the T5 half of the globe. For alpha ($p<0.05$) and beta relative power, the contrary example was seen.

The differentiation was genuinely huge for delta ($p<0.001$), alpha ($p<0.01$), and beta-1 ($p<0.05$) among T1 and T5 patient gatherings. All in all, our single place based investigation discovered that such IoT-based ongoing clinical observing gadgets essentially lessen the intricacy of constant checking and information obtaining processes for a medical services supplier and accordingly give better medical services the board.

The development of critical dangers and controlling components can be improved by supporting the mindfulness. Moreover, it distinguishes the high-hazard factors other than working with the anticipation of strokes. The EEG-based cerebrum PC interface has a promising future in impending a very long time to deflect DALY.

BRIEF DESCRIPTION OF THE DIAGRAM

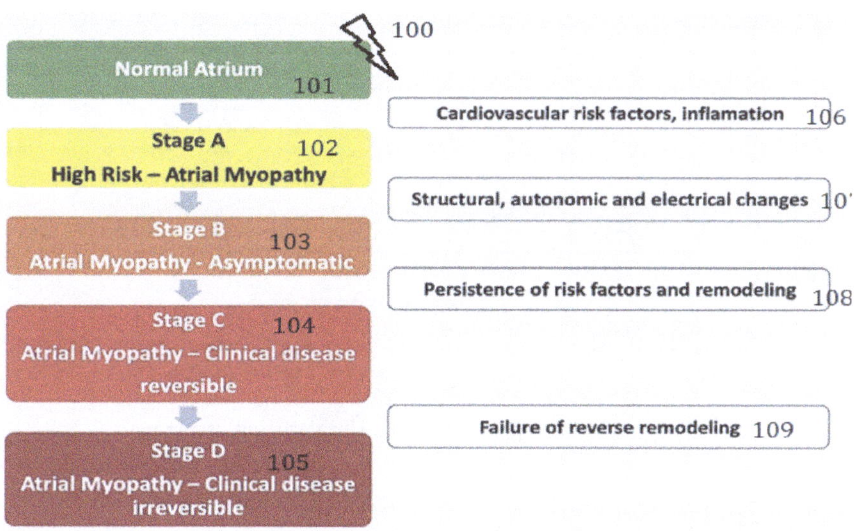

Fig. 5: Independent Predictors of Atrial Fibrillation in Cryptogenic Embolic Stroke Patients using IoT-Based System, Flow Chart.

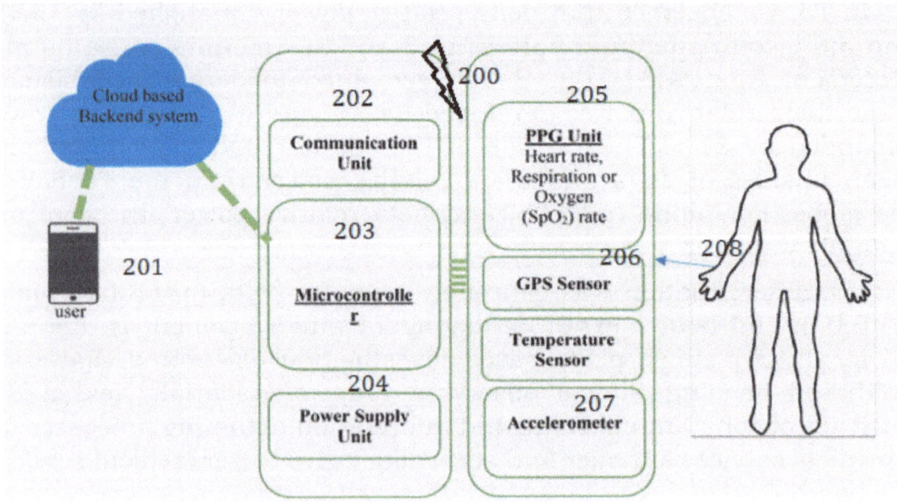

Fig. 6: Independent Predictors of Atrial Fibrillation in Cryptogenic Embolic Stroke Patients using IoT-Based System, Block Diagram.

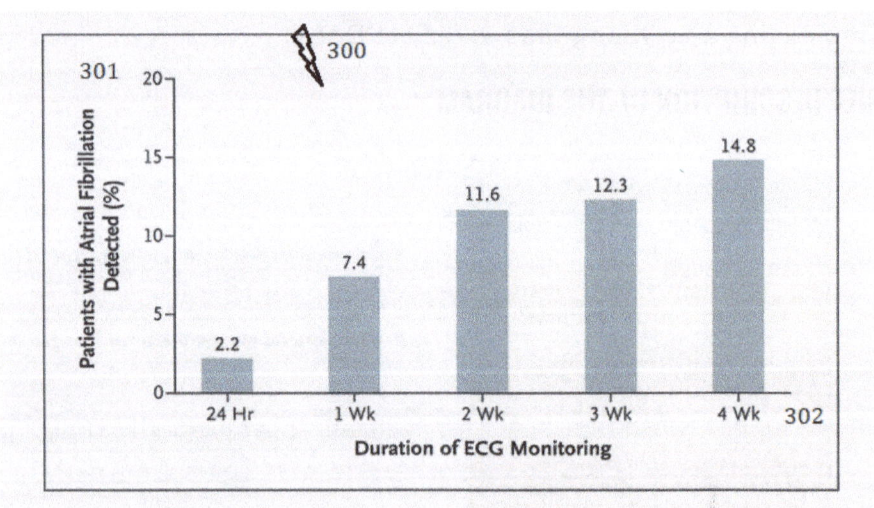

Fig. 7: Independent Predictors of Atrial Fibrillation in Cryptogenic Graph.

DESCRIPTION OF THE INVENTION

In a PPG signal, AF is showed as differing beat to beat spans and heartbeat morphologies. Then again, a typical sinus musicality

(NSR) is conspicuous through consistently divided PPG beats with comparable morphologies between sequential heartbeats. Perceiving an arrhythmia in a PPG sign can at times be trying within the sight of antiquities. Normal wellsprings of ancient rarities are movement and helpless sensor contacts. Antiques can be confused as physiological irregularities. Movement antiques can be distinguished utilizing accelerometric information.

Most current wearable gadgets incorporate accelerometric sensors that action speed increase powers along various spatial headings. It is a generally expected practice to dispose of PPG tainted with an ancient rarity. AF location calculations can be assessed utilizing a few presentation measurements. It is normal for some examinations to report affectability, particularity, and exactness. Affectability is characterized as the likelihood to distinguish genuine AF occasions, while the particularity estimates the extent of real Non-AF cases accurately recognized accordingly.

Exactness is a fair measurement of affectability and explicitness. The exactness of an AF identification calculation is its capacity to separate among AF and Non-AF cases. By and large, exactness is the most well-known revealed metric, alongside the space under the bend (AUC) of the collector working trademark (ROC). A ROC for separating AF versus Non-AF is created by plotting affectability versus (1-explicitness) at various order edges. AUC is a proportion of how well AF cases positioned higher than Non-AF cases.

Since AF has a low commonness it is for the most part necessitated that PPG-based AF finders show high accuracy (positive prescient worth). Rather than announcing the AUC, the region under the accuracy review bend (AUPRC) is an elective metric reasonable for exceptionally imbalanced information (for example low pervasiveness). By and large, any detailed execution metric should consider the low predominance of AF and be assessed on an autonomous test dataset.

Results

In the multivariable model, more seasoned age (chances proportion for 10-year increment, 1.99 [95% CI, 1.78-2.23]; $P<0.001$), hypertension (chances proportion, 1.36 [95% CI, 1.03-1.79]; $P=0.0304$), diabetes (chances proportion, 0.74 [95% CI, 0.56-0.96]; $P=0.022$), and weight list (chances proportion for 5-U increment, 1.29 [95% CI, 1.16-1.43]; $P<0.001$) were free indicators of AF during the review. In an affectability investigation limited to 1117 patients with pattern NT-examining (N-terminal prohormone of cerebrum natriuretic peptide) estimations, just more established age and higher NT-testing were critical free indicators of AF.

Execution of a few distributed prescient models was surveyed, including HAVOC (AF hazard score dependent on hypertension, age ≥75 years, valvular coronary illness, fringe vascular infection, stoutness, congestive cardiovascular breakdown, and coronary vein sickness) and CHA2DS2-VASc (stroke hazard score dependent on congestive cardiovascular breakdown, hypertension, age ≥75 years [doubled], diabetes, past stroke, transient ischemic assault or thromboembolism [doubled], vascular illness, age 65 to 74 years, and sex classification [female]) scores, and higher scores were related with higher paces of creating AF.

Atrial fibrillation was found by short term checking in a remarkable level of patients with cryptogenic stroke. Indicators of mysterious AF in our review populace included PACs and higher TSH levels. Albeit a relationship between low TSH and AF has been grounded, our outcomes recommend that high TSH might be a prescient variable also.

Other than age, the main variable, a few different elements, including hypertension, higher weight list, and absence of diabetes, are autonomous indicators of AF later embolic stroke of unsure source. At the point when benchmark NT-testing was free, just more seasoned age and rise of this biomarker were prescient of resulting AF. Understanding who is at higher danger of creating AF will help with recognizing patients who might profit from more exceptional, long haul heart observing.

CLAIMS

Our Invention “Independent Predictors of Atrial Fibrillation in Cryptogenic Embolic Stroke Patients using IoT-Based System” This was a review cross-sectional investigation of patients determined to have cryptogenic stroke at an exhaustive stroke place and alluded for something like 21 days of drawn out short term telemetry. Telemetry reports and information from the patient’s stroke hospitalization, including imaging studies, electrocardiogram (EKG), echocardiogram, important bodily functions, and research facility information, were surveyed. A modest amount of the 121 patients remembered for the review were determined to have AF dependent on short term telemetry. There was a solid relationship between presence of untimely atrial compressions (PACs) on confirmation EKG and resulting recognition of AF ($P = .004$). Enormous left atrial distance across on echocardiogram was corresponded with AF recognition in guys ($P = .024$). Nonetheless, there was no relationship among AF and other echocardiographic estimations. Thyroid-animating chemical (TSH) levels were essentially higher in patients with cryptogenic stroke having AF (P = .008), with a TSH more noteworthy than 4.20 mIU/L prescient of identification of AF (P < .001). RE-SPECT ESUS was a randomized, controlled preliminary (564 destinations, 42 nations) evaluating dabigatran versus ibuprofen for the avoidance of repetitive stroke in patients with embolic stroke of dubious source. Of 5390 patients enlisted and followed for a middle of 19 months, 403 (7.5%) were found to foster AF revealed as an unfriendly occasion or utilizing heart checking per standard clinical consideration. Univariable and multivariable relapse examinations were performed to characterize indicators of AF.

- According to claim1# the invention is to provide a "Independent Predictors of Atrial Fibrillation in Cryptogenic Embolic Stroke Patients using IoT-Based System" This was a review cross-sectional investigation of patients determined to have cryptogenic stroke at an exhaustive stroke place and alluded for something like 21 days of drawn out short term telemetry.
- According to claim1,2# the invention is to provide a reports and information from the patient's stroke hospitalization, including imaging studies, electrocardiogram (EKG), echocardiogram, important bodily functions, and research facility information, were surveyed.
- According to claim1,2,3# the invention is to provide a modest amount of the 121 patients remembered for the review were determined to have AF dependent on short term telemetry. There was a solid relationship between presence of untimely atrial compressions (PACs) on confirmation EKG and resulting recognition of AF (P = .004). Enormous left atrial distance across on echocardiogram was corresponded with AF recognition in guys (P = .024).
- According to claim1,2,4# the invention is to provide a Nonetheless, there was no relationship among AF and other echocardiographic estimations. Thyroid-animating chemical (TSH) levels were essentially higher in patients with cryptogenic stroke having AF (P = .008), with a TSH more noteworthy than 4.20 mIU/L prescient of identification of AF (P < .001). RE-SPECT ESUS was a randomized, controlled preliminary (564 destinations, 42 nations) evaluating dabigatran.
- According to claim1,2,3,5# the invention is to provide a versus ibuprofen for the avoidance of repetitive stroke in patients with

embolic stroke of dubious source. Of 5390 patients enlisted and followed for a middle of 19 months, 403 (7.5%) were found to foster AF revealed as an unfriendly occasion or utilizing heart checking per standard clinical consideration. Univariable and multivariable relapse examinations were performed to characterize indicators of AF.

REFERENCES

[1] S. J. Park, S. N. Min, H. Lee, M. Subramaniyam, "A driving simulator study: elderly and younger drivers physiological, visual and driving behavior on intersection," in Proc. of the 19th Triennial Congress of the IEA, Melbourne, Australia, pp.1-3, August. 2021.

[2] E. C. Andrews, and S. J. Westerman, "Age difference in simulated driving performance: compensatory process," Accid. Anal. Prev., Vol. 45, pp. 660-668, March, 2021.

[3] J. E. Cohen, "Human population: the next half century," Science, Vol. 302, No. 5648, pp.1172-1175, Nov, 2020.

[4] K. S. Hong, O. Y. Bang, D. W, Kang, et al., "Stroke statistics in Korea: part 1, epidemiology and risk factors: a report from the Korean stroke society and clinical research center for stroke," Journal of Stroke, Vol. 15, No. 1, pp. 2-20, January, 2020.

[5] K. H. Jung, S. H. Lee, B. J. Kim, et al., "Secular trends in ischemic stroke characteristics in a rapidly developed country," Circulation: Cardiovascular Quality and Outcomes, Vol. 5, pp. 327-334, May, 2019.

[6] B. M. Gund, P. N. Jagtap, V. B. Ingale, R. Y. Patil, "Stroke: A Brain Attack," IOSR Journal of Pharmacy, Vol. 3, No. 8, pp. 01-23, September, 2019

[7] The Atlas of Heart Disease and Stroke, WHO, 2016.

[8] D. L. Rimmele, and G. Thomalla, "Wake-up stroke: clinical characteristics, imaging findings, and treatment option–an update," Frontiers in neurology, Vol. 5, Article 35, pp. 1-7, March, 2014.

[9] G. Thomalla, and C. Gerloff, "Treatment concepts for wake-up stroke and stroke with unknown time of symptom onset," Stroke, Vol. 46, No. 9, pp. 2707- 2713, August, 2015.

[10] H. Lee, and S. J. Park, "Quantitative effects of mattress types (comfortable vs. uncomfortable) on sleep quality through polysomnography and skin temperature," Int J Ind Ergon, Vol. 36, No. 11, pp.943–949, November, 2006.

[11] S. J. Park, M. Subramaniyam, M. K. Moon, B. B. Jeon, E. J. Lee, S. H. Han, and C. S. Woo, "Sleep Quality and Skin-Lightening Effects of White Mother Chrysanthemum Aroma," in Industrial Applications of Affective Engineering, Springer International Publishing, Switzerland, April, 2014.

CITATIONS

- Jain, A., Dwivedi, R. K., Alshazly, H., Kumar, A., Bourouis, S., & Kaur, M. (2022). Design and Simulation of Ring Network-on-Chip for Different Configured Nodes. CMC-COMPUTERS MATERIALS & CONTINUA, 71(2), 4085-4100.
- Jain, A., Kumar, A. and Sharma, S., 2015. Comparative Design and Analysis of Mesh, Torus and Ring NoC. Procedia Computer Science, 48, pp.330-337.
- Jain, A., Dwivedi, R., Kumar, A. and Sharma, S., 2017. Scalable design and synthesis of 3D mesh network on chip. In Proceeding of International Conference on Intelligent Communication, Control and Devices (pp. 661-666). Springer, Singapore.
- Jain, A., Gahlot, A.K., Dwivedi, R., Kumar, A. and Sharma, S.K., 2018. Fat Tree NoC Design and Synthesis. In Intelligent Communication, Control and Devices (pp. 1749-1756). Springer, Singapore.
- Jain, A., Dwivedi, R., Kumar, A. and Sharma, S., 2017. Scalable design and synthesis of 3D mesh network on chip. In Proceeding of International Conference on Intelligent Communication, Control and Devices (pp. 661-666). Springer, Singapore.

Chapter - 05

Health Monitoring Using at Mega Microcontroller and IoT

Dr. Raghavendar Raju L

Ph.D.(CSE), Assistant Professor, Dept. of Comouter Science and Engineering
Matrusri Engineering College, Hyderabad-59. lraghavendarraju@matrusri.edu.in

M Venkata Krishna Reddy

Assistant Professor, M.Tech (CSE), (Ph.D.), Dept. Of Computer Science and Engineering,
Chaitanya Bharathi Institute of Technology (A), Gandipet, Hyderabad - 500075
krishnareddy_cse@cbit.ac.in

B Ramana Reddy

Assistant Professor, M.Tech (CSE), (Ph.D.), Dept. Of Computer Science and Engineering,
Chaitanya Bharathi Institute of Technology (A), Gandipet, Hyderabad - 500075
bramanareddy_cse@cbit.ac.in

ABSTRACT

Our Invention "Health Monitoring using at Mega Microcontroller and IOT" The wellbeing observing framework has become famous these days because of uniqueness and expanded utilization in the clinical field. Ordinary many lives are impacted on the grounds that the illnesses are not ideal and appropriately analyzed so we didn't get an opportunity to give clinical assistance. To manage these sorts of circumstances, this framework will assist with observing a patient's sure boundaries and anticipate the patient's condition from time to time. This framework is easy to understand and lessens the human endeavors. This invention provides us with the improvement of a microcontroller based framework for remote heartbeat and temperature observing utilizing Wi-Fi module. By this we can undoubtedly give constant data accessible to numerous clients and can send them alert in basic conditions over web. In India numerous patients are passing on a result of respiratory failures and explanation for this variable is that they are not getting appropriate assistance during the period

***Key:** Health, Monitoring, Mega, Microcontroller, IOT, Wi-Fi module, Framework.*

FIELD

Our Invention is related to a Health Monitoring System using Atmega Microcontroller and IOT- Based. To give them convenient and legitimate assist first we with needing to consistent observing of patient wellbeing. The decent observing framework can be utilized just when the patient is lying on bed and these frameworks are tremendous and just accessible in the medical clinics in ICU.

The framework is produced for home use by patients that are not in a basic condition but rather should be opportune checked by specialist or family. In any basic condition the SMS is ship off the specialist or any relative. So we can undoubtedly save many lives by giving them speedy help.

BACKGROUND

Web of Things (IOT) based shrewd wellbeing observing framework will assist with estimating different wellbeing related boundaries like internal heat level, beat, ECG, pulse and so forth which will assist with anticipating infections. Framework will comprise of a site which will assist with checking patient's wellbeing and furthermore information can be imparted to a specific specialist whenever needed by URL.

This framework can be carried out at our home or advanced age homes to monitor a singular's wellbeing additionally for the everyday wellbeing examination of individuals working at power loads up and public vehicle at the hour of revealing or exit. This framework can likewise be useful to screen the strength of excavators and individuals working at the shipper naval force and to screen the patients in the emergency clinic who are held under perception.

A remote sensor organization (WSN) is a remote organization which comprises of fundamentally circulated independent gadgets that utilization sensors to screen physical or ecological conditions.

These independent gadgets, or hubs, join with switches and a passage to make a WSN framework. Sensor networks are the way to social occasion the data required by brilliant conditions, regardless of whether in structures, utilities, ventures, home, ships, transportation frameworks computerization, or somewhere else.

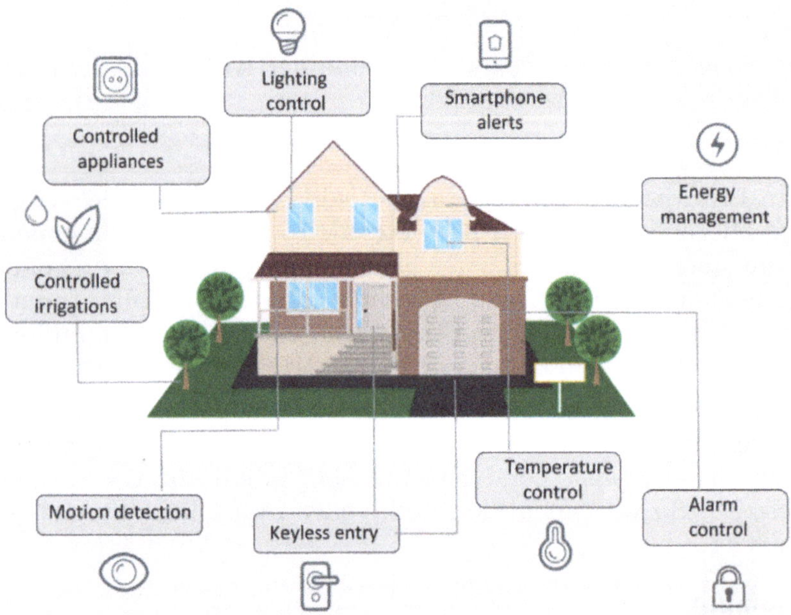

Fig. 1: Health Monitoring System using Atmega Microcontroller and IOT- Based Flow Chart.

Late psychological oppressor and hit and run combat counter measures require dispersed organizations of sensors that can be extended and have self-organizing capacities. In such applications, running wires or cabling is typically unfeasible. A sensor network is required that is quick and simple to introduce and keep up with.

The shrewd entryway is intended to empower WSN and public correspondence organizations to get to one another with consistent internetworking. In this plan, the door comprises of focal control unit, data set (DB), WSN module, WLAN AP, and GSM module. Construction of Gateway The conveyed estimation hubs impart remotely to a focal door, which gives an association with the wired reality where the information can be gather, process, dissect and introduce your estimation information.

To expand distance and dependability in a remote sensor organization, you can utilize switches to build an additional a correspondence interface between end hubs and the door. The passage incorporates three outside correspondence modules (ECM): WSN Module, WLAN AP, and GSM Module. WSN module, on one hand, is essentially utilized for getting information bundles from the center of the WSN; then again, it is utilized to send orders to the WSN or explicit sensor hubs.

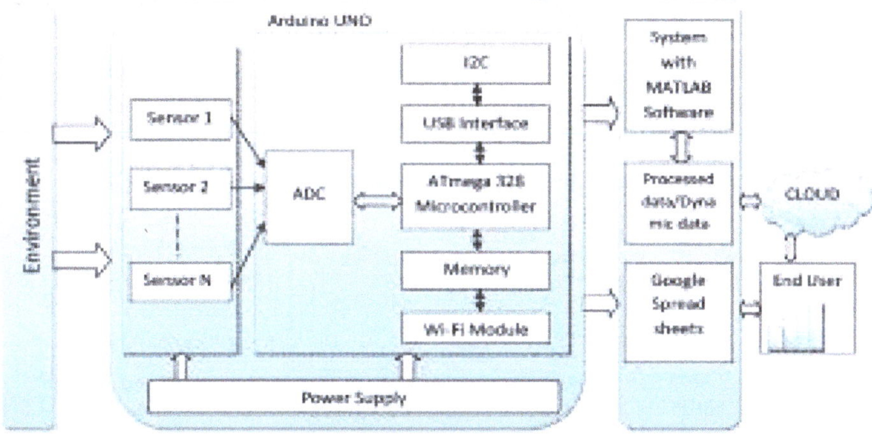

Fig. 2: Health Monitoring System using Atmega Microcontroller and IOT- Based Flow Chart.

Web of Things (IoT) is the arising worldview, which contains colossal measure of savvy article and brilliant gadgets associated with the web for speaking with one another. These shrewd gadgets are utilized to gather temperature, pulse, sugar level and so forth, which are utilized to assess the ailment of the patient. IoT gadgets are utilized in many fields which make the clients' everyday life more agreeable.

Conveying the gathered data to the specialist, settling on precise choice on the information gathered and telling the patient is the difficult undertaking in the IoT. In this invention, the engineering of the Patient Health Monitoring System (PHMS) utilizing IoT gadgets is proposed to gather the necessary boundaries and assess the information acquired from the IoT gadgets. PHMS additionally advises the patient with conceivable careful steps to be drilled by them.

This framework recommends the patient with clinical consideration and subsequent stage to be continued in the event of basic circumstance. The PHMS framework is assessed for specific boundaries and the choices made on the information acquired from the source are accepted to assess the framework. The proposed framework is intended for pulse, Respiratory rate, kidney stone and kidney disappointment recognition.

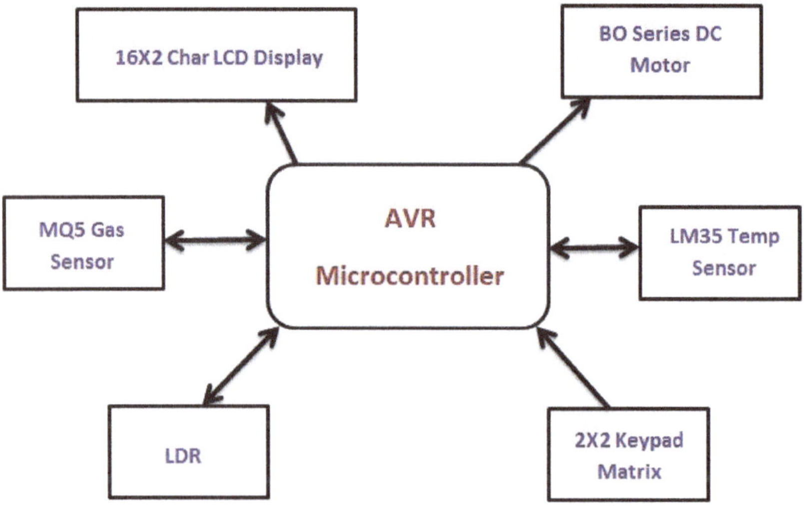

Fig. 3: Health Monitoring System using Atmega Microcontroller and IOT- Based Flow Chart.

OBJECTIVES

The objective of the invention is to provide a "Health Monitoring System using Atmega Microcontroller and IOT- Based" The wellbeing observing framework has become famous these days because of uniqueness and expanded utilization in the clinical field.

The other objective of the invention is to provide a Ordinary many lives are impacted on the grounds that the illnesses are not ideal and appropriately analyzed so we didn't get an opportunity to give clinical assistance.

The other objective of the invention is to provide a manage these sorts of circumstances, this framework will assist with observing a patient's sure boundaries and anticipate the patient's condition from time to time. This framework is easy to understand and lessens the human endeavors.

The other objective of the invention is to invention provides us with the improvement of a microcontroller based framework for remote heartbeat and temperature observing utilizing Wi-Fi module. By this we can undoubtedly give constant data accessible to numerous clients and can send them alert in basic conditions over web.

The other objective of the invention is to numerous patients are passing on a result of respiratory failures and explanation for this variable is that they are not getting appropriate assistance during the period.

SUMMARY

First System Here, specialist planned wellbeing checking framework utilizing ATmega8 microcontroller with Wireless Body Area Sensor Network (WBASN). In this work, the sensors which are utilized here are Temperature sensor, Blood pressure sensor, Heart thump sensor.

These sensors are put on the human body which assists with observing the ailment without upsetting the day by day timetable of the patient and these wellbeing related boundaries are then sent to doctor's waiter utilizing long reach remote innovation GSM.

Wellbeing checking framework comprises of sensors, microcontroller, LCD show and GSM modem to send or get wellbeing related information to or from the specialist. Likewise, at the clinic a similar GSM modem is utilized. Thus, GSM modem helps in the foundation of an organization between persisting's server and specialist's server. LCD (Liquid Crystal Display) show is given to show the moment result to the patient.

Here specialist involved LM34 as temperature sensor, IR LED and red LED is utilized for pulse observing and Pressure transducer or the sensor in light of piezo-electrical material is utilized to gauge the systolic BP and diastolic BP. Microcontroller peruses information as given by the temperature sensor, circulatory strain sensor and pulse sensor and handling it gives the result as computerized and it gets straightforwardly shown on LCD or it gets communicated to the specialist's server through GSM modem.

This framework gives definite and moment results with high precision which gets straightforwardly shown on LCD. It takes max 4-5 sec to screen the specialist's server utilizing GSM remote innovation .This framework invests in some opportunity to realize the medical issue of the patient and afterward conveys the report to the specialist.

The Internet of things is the internetworking of actual gadgets, e-wellbeing screen, vehicles, structures, and different things installed with hardware, programming, sensors, actuators, and organization availability that empower these items to gather and trade information. In 2013 the Global Standards Initiative on Internet of Things (IoT-GSI) characterized the IoT as "the framework of the data society."

The IoT permits objects to be detected or potentially controlled from a distance across existing organization foundation, setting out open doors for more straightforward joining of the actual world into PC based or advanced cell based frameworks, and bringing about better productivity, exactness and financial advantage notwithstanding diminished human mediation.

DIAGRAM

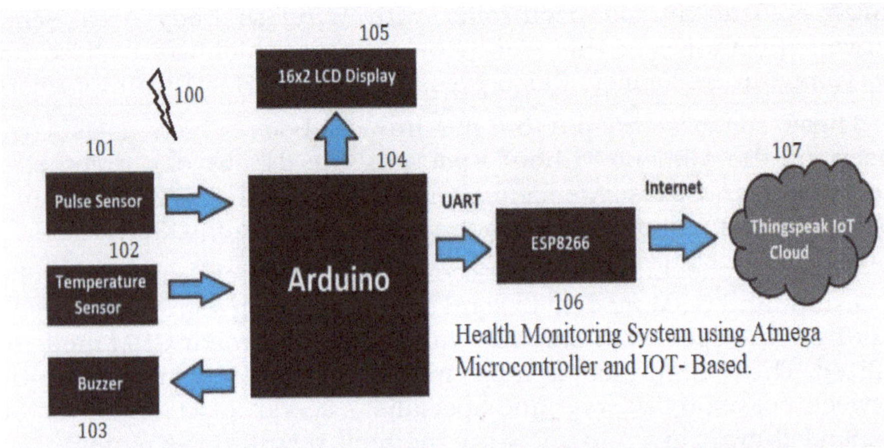

Fig. 4: Health Monitoring System using Atmega Microcontroller and IOT- Based Flow Chart.

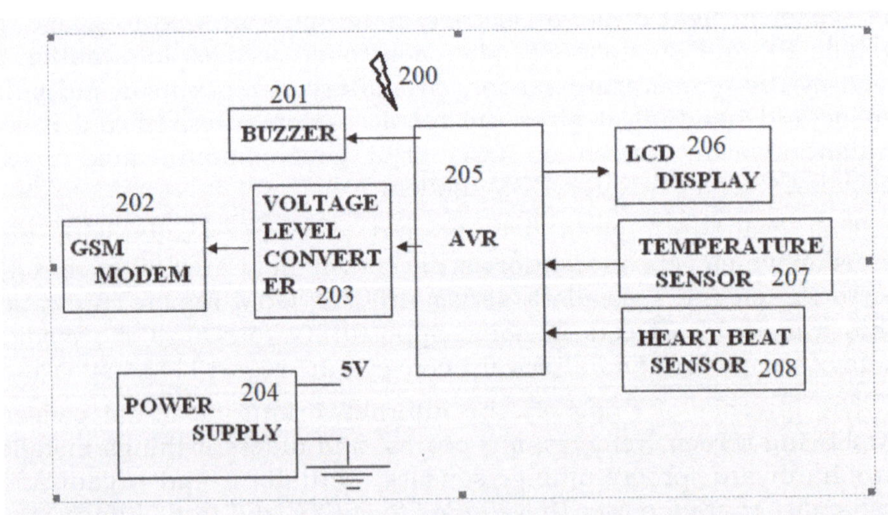

Fig. 5: Health Monitoring System using Atmega Microcontroller and IOT- Based Block Diagram.

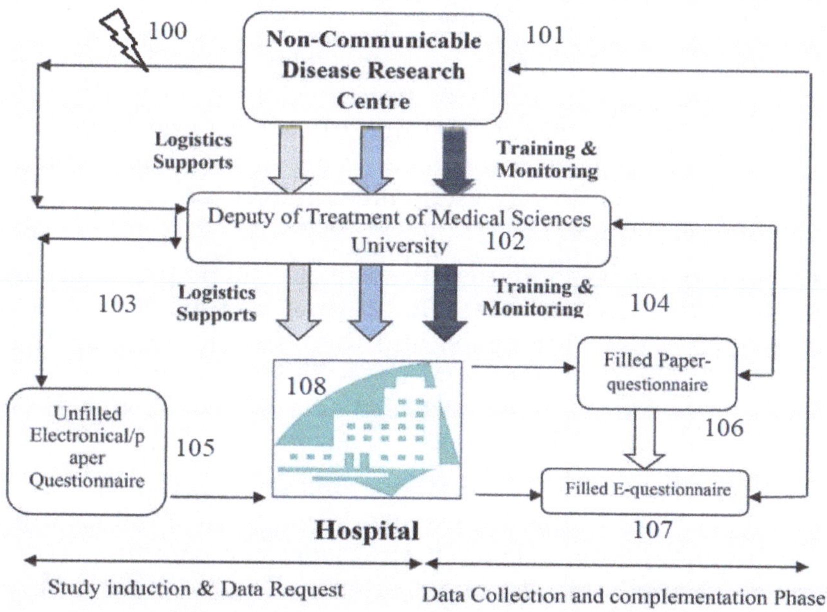

Fig. 6: Health Monitoring System using Atmega Microcontroller and IOT- Based Block Diagram

INVENTION

With the advancement of world, Health checking framework is utilized each field like emergency clinic, home consideration unit, sports. This wellbeing observing framework use for narrative illnesses patients who have every day examination. Ordinarily it is hard to keep track on anomalies in heartbeat count for patient itself physically. The normal heartbeat each moment for 25-year old reaches between 140-170 bpm while for a 60-year old it is around between 115-140 bpm and internal heat level is 37degree Celsius or 98.6 Fahrenheit.

Patients are not knowledgeable with manual treatment which specialists typically use for following the count of heartbeat. There are different instruments accessible in market to keep track on interior body changes. In any case, there are many cutoff points in support part because of their weighty expense, size of instruments and portability of patients.

Heart Beat Monitor with microcontroller ATMEGA16 (AVR). Heart Beat Monitoring circuit utilizing LM358 The utilization of microcontroller is in each field even we can involve it in the plan and creation of biomedical types of gear.

A model is given here. The microcontroller AT89S51 (8051) is here utilized for advancement of a heartbeat checking framework. By setting your finger in the middle, a LED and photograph resistance, person can undoubtedly recognize the beats of heart, the simple voltages are additionally handled with a functional enhancer LM 358, this chip has two inherent Op-Amps.

The TTL beats or computerized beat are then feed to the outside hinder of AVR microcontroller. By utilizing a product counter in the code, they can count the beats, and the outcome the cycle is shown on a LCD (2 line 16 characters). HB = 5184/t Where t is normal of time delay between 2 sequential heartbeats here utilization of initial 5 heartbeats for computation of HB.

TEMPERATURE SENSOR (LM35) - This series comprise of accuracy coordinated circuit temperature sensors, whose yield voltage is directly corresponding to the Celsius (Centigrade) temperature. The LM35 doesn't need any outer alignment or managing to give exactness's of (+-)1/4 degree Celsius at room temperature and (+-)3/4 degree Celsius over a full - 55 to +150-degree Celsius temperature range. Less to works from 4 to 30 volt. Under 60uA current channel.

According to invention work, wellbeing checking framework configuration depends on specialist thought that meets to the patients need. According to thought of customary framework, this framework still being used from their assembling yet it is exceptionally cumbersome to deal with separately and size and cost are likewise more contrasted with the development framework and furthermore it take more than 1minute for getting the specific outcome.

According to thought of advance framework, every framework has its own benefit. Every wellbeing checking framework has different detail according to patient's prerequisite. This framework gives more clinical instrument office on single framework on-chip contrast with traditional framework.

This framework takes under 1 moment to work out outcome connected with ailment. Size likewise lessens contrasted with the customary framework due to incorporation of number of clinical instrument on single chip. Along these lines, size, cost and intricacy additionally diminish.

As thought of microcontroller like Atmega (AVR), PIC regulator there is need to associate outside fringe for signal molding. Subsequently, size, cost and adaptability are more. Subsequently, as outer fringe builds cost and size additionally increments. Specialists planned wellbeing observing framework according to patient's prerequisite.

In view of remote information transmission over web, wellbeing related information will be ship off specialist's PC or on his versatile. In this way, need to go emergency clinic without fail and sending back rub to the specialist gets quick cure connected with the ailment.

In Iris Image Analysis strategy kidney status is recognized. The fundamental thought is to consolidate IRIDOLOGY and Image handling with its logical motivations to open up another examination field and medical care indicative strategy. Carrying out this strategy basically requires unique picture handling where iris include extraction assumes an essential part.

Input iris picture is taken as information and given it to pre-handling stage where picture quality is upgrading with fundamental change. Pre-handled iris picture is extricated and divided. Well acknowledgment technique is utilized to identify kidney stone and kidney disappointment. Matlab picture handling tool kit is utilized to carry out this task.

Heart beat sensor: Heart thump sensor is intended to give computerized result of hotness thump when a finger is put on it. At the point when the heart beat identifier is working, the thump LED streaks as one with every heartbeat. This computerized result can be associated with microcontroller straightforwardly to gauge the Beats each Minute (BPM) rate. It deals with the rule of light regulation by blood course through finger at each heartbeat.

Respiratory Sensor

The Respiration Sensor is utilized to screen stomach or thoracic breathing, in biofeedback applications, for example, stress the executives and unwinding preparing. The Respiration Sensor is typically positioned in the stomach region, with the focal piece of the sensor simply over the navel.

The sensor ought to be set tight to the point of forestalling deficiency of pressure. An individual's respiratory rate is the quantity of breaths you take each moment. The ordinary breath rate for a grown-up very still is 12 to 20 breaths each moment. A breath rate under 12 or more than 25 breaths each moment while resting is viewed as unusual.

Research Outcome

Our Invention "Health Monitoring System using Atmega Microcontroller and IOT- Based" The wellbeing observing framework has become famous these days because of uniqueness and expanded utilization in the clinical field. Ordinary many lives are impacted on the grounds that the illnesses are not ideal and appropriately analyzed so we didn't get an opportunity to give clinical assistance. To manage these sorts of circumstances, this framework will assist with observing a patient's sure boundaries and anticipate the patient's condition from time to time. This framework is easy to understand and lessens the human endeavors. This invention provides us with the improvement of a microcontroller based framework for remote heartbeat and temperature observing utilizing Wi-Fi module. By this we can undoubtedly give constant data accessible to numerous clients and can send them alert in basic conditions over web. In India numerous patients are passing on a result of respiratory failures and explanation for this variable is that they are not getting appropriate assistance during the period.

- According to claim1# the invention is to a Our Invention "Health Monitoring System using Atmega Microcontroller and IOT-Based" The wellbeing observing framework has become famous these days because of uniqueness and expanded utilization in the clinical field.
- According to claim1,2# the invention is to a Ordinary many lives are impacted on the grounds that the illnesses are not ideal and appropriately analyzed so we didn't get an opportunity to give clinical assistance.
- According to claim1,2,3# the invention is to a manage these sorts of circumstances, this framework will assist with observing a patient's sure boundaries and anticipate the patient's condition from time to time. This framework is easy to understand and lessens the human endeavors.
- According to claim1,2,3,4# the invention is to a invention provides us with the improvement of a microcontroller based framework for remote heartbeat and temperature observing utilizing Wi-Fi module. By this we can undoubtedly give constant data accessible to numerous clients and can send them alert in basic conditions over web.
- According to claim1,2,3,4# the invention is to a numerous patients are passing on a result of respiratory failures and explanation for this variable is that they are not getting appropriate assistance during the period.

REFERENCES

[1] Gulraiz J. Joyia, Rao M. Liaqat, Aftab Farooq, and Saad Rehman, Internet of Medical Things (IOMT): Applications, Benefits and Future Challenges in Healthcare Domain, Journal of Communications Vol. 12, No. 4, April 2022.

[2] Shubham Banka, Isha Madan and S.S. Saranya, Smart Healthcare Monitoring using IoT. International Journal of Applied Engineering Research ISSN 0973-4562 Volume 13, Number 15, pp. 11984-11989, 2020.

[3] K. Perumal, M. Manohar, A Survey on Internet of Things: Case Studies, Applications, and Future Directions, In Internet of Things: Novel Advances and Envisioned Applications, Springer International Publishing, (2019) 281-297.

[4] S.M. Riazulislam, Daehankwak, M.H.K.M.H., Kwak, K.S.: The Internet of Things for Health Care: A Comprehensive Survey. In: IEEE Access (2020).

[5] P. Rizwan, K. Suresh. Design and development of low investment smart hospital using Internet of things through innovative approaches, Biomedical Research. 28(11) (2017).

CITATIONS

- Jain, A., Kumar, A. and Sharma, S., 2015. Comparative Design and Analysis of Mesh, Torus and Ring NoC. Procedia Computer Science, 48, pp.330-337.
- Jain, A., Dwivedi, R., Kumar, A. and Sharma, S., 2017. Scalable design and synthesis of 3D mesh network on chip. In Proceeding of International Conference on Intelligent Communication, Control and Devices (pp. 661-666). Springer, Singapore.
- Jain, A., Gahlot, A.K., Dwivedi, R., Kumar, A. and Sharma, S.K., 2018. Fat Tree NoC Design and Synthesis. In Intelligent Communication, Control and Devices (pp. 1749-1756). Springer, Singapore.
- Gupta, N., Jain, A., Vaisla, K.S., Kumar, A. and Kumar, R., 2021. Performance analysis of DSDV and OLSR wireless sensor network routing protocols using FPGA hardware and machine learning. Multimedia Tools and Applications, 80(14), pp.22301-22319.

Chapter - 06

VoWiFi Cell Capacity IEEE 802.11ax for VBR Traffic using IoT

Dr. Muddamalla Naresh

Assistant Professor, ECE Department, Matrusri Engineering College, Hyderbad,Telanaga
nareshmuddamalla@matrusri.edu.in

Thanam Pullaiah

Associate Professor, ECE Department, Vignan's Institute of Management and Technology for Women, Hyderabad, Telangana. thanam.tp@gmail.com

ABSTRACT

Apart from mobile cellular networks, IEEE 802.11based wireless local area networks (WLANs) represent the most widely deployed wireless networking technology. With the migration of critical applications onto data networks, and the emergence of multimedia applications such as digital audio/video and multimedia games, the success of IEEE 802.11 depends critically on its ability to provide quality of service (QoS). A lot of research has focused on equipping IEEE 802.11 WLANs with features to support QoS. In this survey, we provide an overview of these techniques. We discuss the QoS features incorporated by the IEEE 802.11 standard at both physical (PHY) and media access control (MAC) layers, as well as other higher-layer proposals. We also focus on how the new architectural developments of softwaredefined networking (SDN) and cloud networking can be used to facilitate QoS provisioning in IEEE 802.11-based networks. We conclude this paper by identifying some open research issues for future consideration.

Key: *VoWiFi, Cell, Capacity, IEEE 802.11,ax VBR ,Traffic, IoT, media access control.*

INTRODUCTION

The IEEE 802.11 wireless local area networking (WLAN) standard defines one of the most widely deployed wireless technologies in the world. The popularity of wireless networking is driven by the ubiquity of portable mobile hand-held devices, and the convenience of untethered communications. With the increasing deployment of multimedia content on the Internet—such as digital video, voice over IP (VoIP), videoconferencing, and multi-player networked games—along with the deployment of time-sensitive critical applications, there is a strong motivation to develop QoS features to meet the more stringent performance requirements [1].

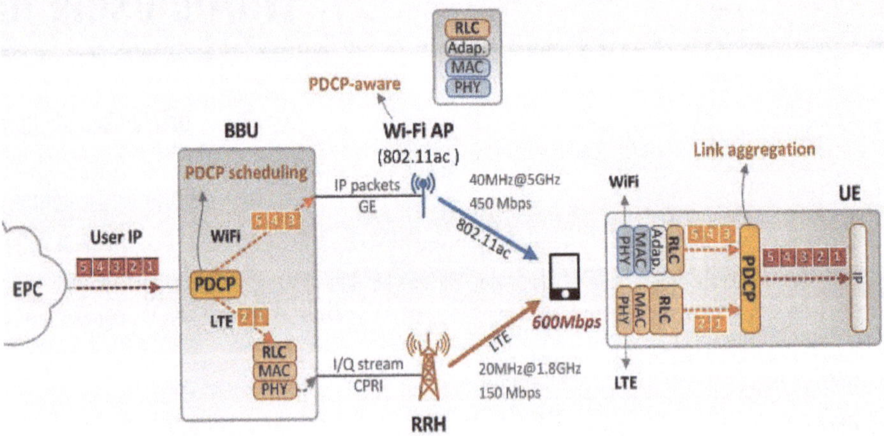

Fig. 1: VoWiFi Cell Capacity IEEE 802.11ax for VBR Traffic using IoT

While the Internet and data networking models of the IEEE 802.11 WLAN technology, which are based on the datagram delivery model of IP, provide simple, adaptive and fault resilient network, they are ill-suited to QoS provisioning. The underlying datagram model of IP is a best-effort service— i.e., while the network tries to deliver packet to the destination correctly without any packet losses, it makes no guarantees. Multimedia applications, in particular, need stronger guarantees about the minimum throughput and maximum latency to work satisfactorily. An expensive solution for ensuring QoS is to overprovision. Most of the Internet QoS effort has focused on how to get a network with less capacity meet application

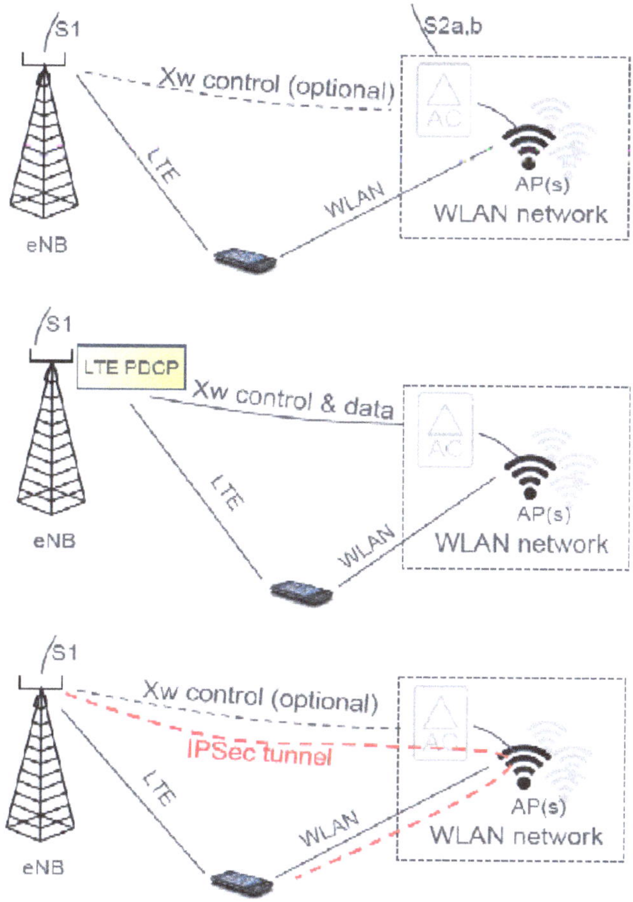

Fig. 2: VoWiFi Cell Capacity IEEE 802.11ax for VBR Traffic using IoT

Service Differentiation

Service differentiation is used to support multiple services with diverse requirements—such as interactive delay-sensitive services along with elastic delay-tolerant file transfer services [2][4]. The overprovisioning of network resources is not always possible in radio networks, thus making service differentiation an integral component of most QoS-based solutions. In service differentiation, several parameters (e.g., packet deadline) can be modified to define how a flow should access the wireless medium [4]. A variety of services can be provided by the use of simple network parameters deployed in network nodes, and these services can be classified according to a large number of characteristics [25]. The QoS of the system is enhanced by differentiating the priority of each host and offering them different levels of QoS parameters.

Admission Control One Way of Supporting QoS is through Admission Control

In which new sessions are allowed onto the network only if sufficient resources are available to provide service to the new and existing sessions [30]. The interest in the field of admission control, has been driven by the idea that regulation of incoming traffic flows prevents network congestion, and helps in ensuring QoS. Call Admission Control (CAC) is a traffic management system employed in public switched telephone networks (PSTNs) [3][1]. The CAC scheme is easy to implement because of the homogeneous environment.

However, the present networking environment of homogeneous network is not preserved. Hence, the admission control function is more challenging in heterogeneous networks (e.g., joint WLANs and IP networks) [3][2]. A new flow request is admitted only if the ongoing flows are not negatively affected. The common parameters used for admission control are peak bandwidth requirement and the average rate [9]. Therefore, the task of admission control is to maximize resource utilization in the network, and to control the amount of traffic to achieve the predefined performance objectives of the current flows.

Congestion Control Congestion control in the modern Internet is typically performed using the TCP protocol [3][4]. Congestion in a network may occur if the number of packets sent to the network is greater than the number of packets a network can handle. Congestion control refers to the techniques to control the congestion level and keep the load below the capacity.

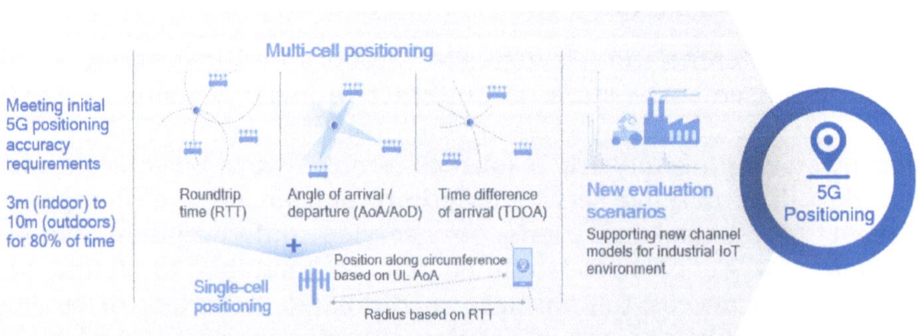

Fig. 3: VoWiFi Cell Capacity IEEE 802.11ax for VBR Traffic using IoT

In the QoS-integrated services, the congestion control mechanism should be different for different kinds of sources: e.g., file transfer/ email is different from real-time voice/video applications [3][5]. The QoS enabled routers provide services to certain flows based on their requirements. Congestion control helps to provide priority differentiation of flows by servicing queues in different manners (e.g., the order in which the flows are serviced).

Scheduling

Scheduling is the key to share network resources fairly among users in a network, and it provides service guarantees to time-critical applications. The scheduler first decides the order of requests to be served, and then it manages the queues of these awaiting requests. The scheduling scheme is important for the networks because there are two types of applications. One is insensitive to the performance that users receive from the network, and the other has a strict bound on the performance.

The scheduling can provide different services to the flows using parameters such as different bandwidths— by serving only a single flow at a particular interval; different mean delays—according to the level of priority defined for the flow; and different loss rates—by assigning more or fewer buffers to the flows [3][6]. The scheduling mechanism adopted in the IEEE 802.11 standard is explained in detail later in Section III-C.

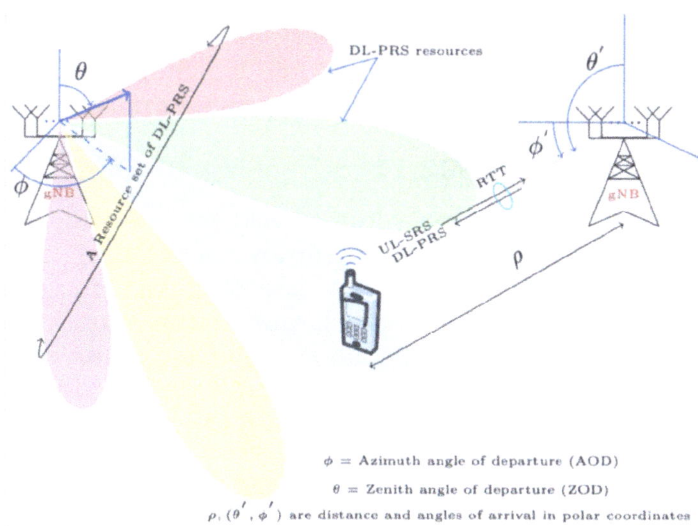

Fig. 4: VoWiFi Cell Capacity IEEE 802.11ax for VBR Traffic using IoT

Standards in the IEEE 802 project target the PHY layer and the MAC layer. While IEEE 802.3 defines the PHY and MAC layers for wired LANs; the prominent IEEE 802.11 standard, which is the focus. The first IEEE 802.11 specification was published in 1997, and it has undergone numerous subsequent amendments. The IEEE 802.11 working group has various task groups focusing on a myriad of niche concerns with an elaborated description of the IEEE 802.11 universe provided in [38]. The focus of the various task groups is summarized in tabulated form in Tables II and III. Providing QoS services in applications using traditional IEEE 802.11 standards is difficult since they provide no explicit mechanisms for service differentiation.

Various parameters extracted from the general traffic layout are used to ensure acceptable QoS in these networks, such as goodput— which is the measure of packet arrival rate during a fixed period of time; load level—which indicates the usage of a medium on per time basis; and available bandwidth—which measures the rate at which new flows can send traffic without affecting the existing flows in the network [3][9] [4][0]. The legacy standards only have the basic distributed coordination function (DCF) and the optional point coordination function (PCF) enhancements, such as collision avoidance and a first in f irst out (FIFO) scheduler [4][1].

Even with the contention-free PCF, the QoS problem could not be solved. Consequently, the services provided to the users do not have optimal performances for various applications including audio and video applications during heavy network loads [4][2]. In Section III-D, we present the IEEE 802.11e standard, which provides MAC layer enhancements for QoS, that incorporates traffic priority.

QoS Features for IEEE 802.11

In general, the major techniques used for ensuring QoS at the MAC layer include admission control and scheduling. In the IEEE 802.11 standard, the MAC layer provides the functionality of addressing, framing, reliability check, and access coordination to the wireless medium [5][0]. The MAC layer with QoS enhancements aims to provide the network with a much reduced overhead, segregating frames on the priority basis, and keeping the collisions to the least possible level.

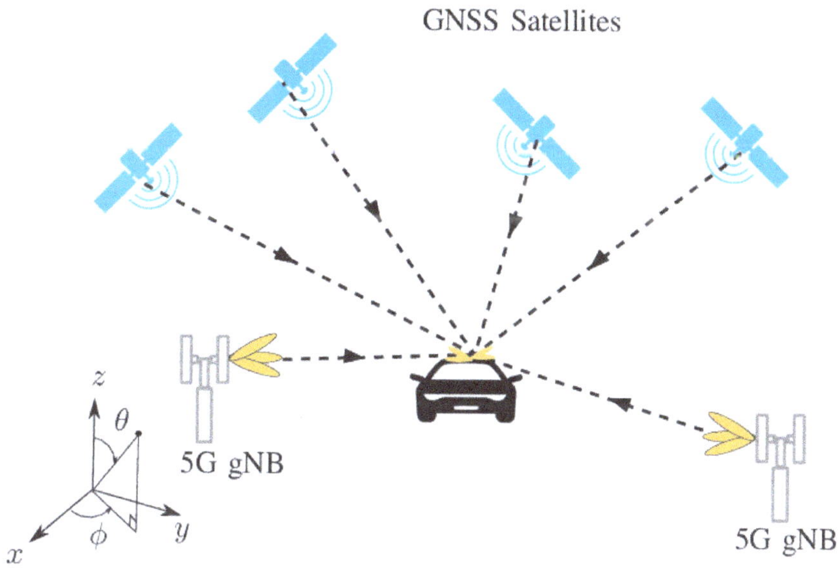

Fig. 5: VoWiFi Cell Capacity IEEE 802.11ax for VBR Traffic using IoT

This section describes the techniques implemented in the legacy IEEE 802.11 standard. The rest of this subsection presents a description of the QoS-focused IEEE 802.11e standard. The wireless systems can be configured in two different modes in the IEEE 802.11 architecture: (i) the ad-hoc mode, and (ii) the infrastructure mode. The infrastructure mode has multiple stations that can communicate with each other through an access point (AP), where the APs have connectivity with a wired network at the backend. While in the adhoc mode, the stations can communicate directly with each other without any intervening access points or a backend wired network. The IEEE 802.11 has two medium access coordination functions, namely the DCF and the PCF

CONCLUSION

It is anticipated that the bulk of access technologies in the near future will be wireless. With emerging standards such as 5G, and the convergence of the telecom and Internet industries on IP-based technologies, the ability to provide high QoS has become paramount. In this paper, we have surveyed various QoS enhancement techniques proposed for IEEE 802.11based wireless networks with our discussion encompassing both classical techniques as well as proposals for new and emerging architectures such as SDN and cloud- networks.

We have classified these techniques using different criteria. We have discussed QoS solutions that have been proposed for various TCP/IP layers along with a discussion on cross-layered protocols. We have also highlighted open research challenges and directions for future work.

REFERENCES

[1] N. Baghaei and R. Hunt, "Review of quality of service performance in wireless LANs and 3G multimedia application services," Computer Communications, vol. 27, no. 17, pp. 1684–1692, 2022.

[2] M. Gast, 802.11 wireless networks: the definitive guide. " O'Reilly Media, Inc.", 2022

[3] Z. Wang, Internet QoS: architectures and mechanisms for quality of service. Morgan Kaufmann, 2021.

[4] D. Chen and P. K. Varshney, "QoS support in wireless sensor networks: Asurvey." in International Conference on Wireless Networks, vol. 233, 2019, pp. 1–7.

[5] P. S. Mogre, M. Hollick, and R. Steinmetz, "QoS in wireless mesh networks: challenges, pitfalls, and roadmap to its realization," in International workshop on Network and Operating Systems Support for Digital Audio & Video (NOSSDAV), 2017.

[6] H. Zhu, M. Li, I. Chlamtac, and B. Prabhakaran, "A survey of quality of service in IEEE 802.11 networks," Wireless Communications, IEEE, vol. 11, no. 4, pp. 6–14, 2014.

[7] A. Lindgren, A. Almquist, and O. Schel´en, "Evaluation of quality of service schemes for IEEE 802.11 wireless lans," in Local Computer Networks, 2001. Proceedings. LCN 2001. 26th Annual IEEE Conference on. IEEE, 2020, pp. 348–351.

CITATIONS

- Jain, J., Jain, A., Srivastava, S. K., Verma, C., Raboaca, M. S., & Illés, Z. (2022). Improved Security of E-Healthcare Images Using Hybridized Robust Zero-Watermarking and Hyper-Chaotic System along with RSA. Mathematics, 10(7), 1071.
- Jain, J., & Jain, A. (2022). Securing E-Healthcare Images Using an Efficient Image Encryption Model. Scientific Programming, 2022.
- Kumar, S., Shailu, A., Jain, A., & Moparthi, N. R. (2022). Enhanced Method of Object Tracing Using Extended Kalman Filter via Binary Search Algorithm. Journal of Information Technology Management, 14(Security and Resource Management challenges for Internet of Things), 180-199.
- Jain, A., Dwivedi, R. K., Alshazly, H., Kumar, A., Bourouis, S., & Kaur, M. (2022). Design and Simulation of Ring Network-on-Chip for Different Configured Nodes. CMC-COMPUTERS MATERIALS & CONTINUA, 71(2), 4085-4100.
- Sharma, S.K., Jain, A., Gupta, K., Prasad, D. and Singh, V., 2019. An internal schematic view and simulation of major diagonal mesh network-on-chip. Journal of Computational and Theoretical Nanoscience, 16(10), pp.4412-4417.

Chapter - 07

QR-Code Based Digital Marketing Using IoT

Satender Kumar

Quantum University, Dehradun Highway, Mandawar, Roorkee, Uttarakhand 247167, India.

Chandani Sharma

Quantum University, Dehradun Highway, Mandawar, Roorkee, Uttarakhand 247167, India.

Dipti Deshpande

Assistant Professor, ISB&M College of Commerce
diptiyogeshdeshpande@gmail.com

Manish Sharma

Quantum University, Dehradun Highway, Mandawar, Roorkee, Uttarakhand 247167, India.

ABSTRACT

Our Invention " |Customer| Preferences on QR-Code Notification Based Digital Marketing" is a || that is not the source or originator of a |QR code| is empowered to give a shopper a determination of related substitute substance instead of satisfying recognized in the |QR code.| A strategy is |carried| out on a PC |framework| to get extra substance data addressed by a QR code removed by a buyer's |QR peruse| gadget from a |media| show containing the |QR code| and also the QR code is |produced| by a wellspring of the media. An area of substitute substance from a hostage climate connected with the extra substance data is recognized. The area of the substitute substance might be |recognized| dependent on the decoded data from the QR code. The area of the |substitute| substance is sent to the QR peruse gadget. The point of this invention is to decide client inclinations towards various procedures of Digital Marketing used to advance offer of electronic gadgets. The interest in this sort of items is |relied upon| to increment later on, thusly it is fundamental for organizations to track down the most effective way to stand apart from the opposition |Some| |what| recently, and the advancement of current innovation has made Digital Marketing the main method of association among organizations and their clients. Advanced Marketing apparatuses offer wide scope of chances for organizations to draw in and hold buyers however| they need to know how their clients assess their advertising endeavor. To analyze which promoting procedures buyers like in this invention we utilize Choice Based Conjoint Analysis (CBC). This strategy requests the respondents to pick between various mixes from promoting components, which addresses common decision-making issue they face in day-to-day existence. |Likewise||CBC| offers us the chance to examine possible connections between given components. Knowing which blend is-viewed as the most favoured one could assist organizations with distributing their assets and make promoting procedure the s that will guarantee constructive outcomes on their business execution.

Key: *QR-Code, Digital, Marketing, IoT, Choice Based, Conjoint, Analysis.*

FIELD

Our Invention is related to a Customer Preferences on QR-Code Notification Based Digital Marketing.

BACKGROUND

The investigation of Quick Response (QR) codes in advertising can be set with regards to portable innovation. Portable innovation has affected the manner in which we associate, impart, work and invest our recreation energy. The innovation is currently all around acknowledged and utilized.

Be that as it may, organizations must comprehend customer reception conduct to this innovation before any tremendous venture choice is made. Buyers see portable advertising interchanges to be differently disturbing protection attack and nosy.

Considering these worries, numerous specialists have proposed consent based showcasing where clients have command over the messages and content they get. QR codes are one innovation of premium with respect to pull advertising correspondence. As of late, there have been numerous helpful ways of adding virtual data to an actual item, to give valuable substance frequently at the period of scarcity.

Versatile labeling offers the chance of such a simple access by connecting portable administrations to actual articles. Shoppers generally need simple and prompt admittance to required data and

QR codes are currently being utilized to make that conceivable. Many organizations have made progress whiles others have fizzled.

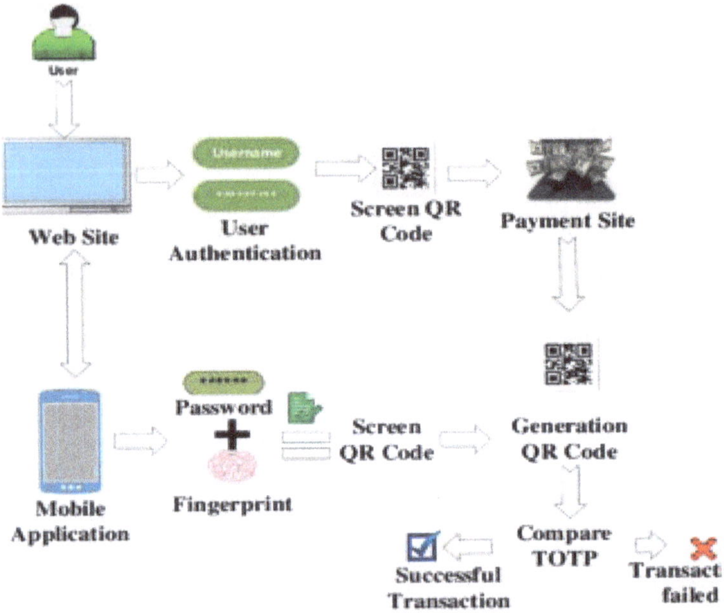

Fig. 1: QR-Code Notification Based Digital Marketing, Process

Speedy reaction (QR) codes are two-layered and machine-meaningful grid standardized tags initially created and planned in 1994 by Denso Wave Incorporated with the end goal of precise and quick stock checks.1 Conventional one-layered scanner tags can store data just in the even way, while QR codes can store data both in an upward direction and evenly. So more great and enormous data can be put away in QR codes than in one-layered customary standardized identifications.

Fig. 2: QR-Code

The put away data can be encoded as a URL, message, or other different sorts of information, which can be effortlessly perused by the cameras of versatile device.3 With the expanding fame of camera-empowered cell phones, the clients of QR codes increment quickly. Japanese publicists and advertisers quickly accepted these codes as showcasing apparatuses in an expansive scope of promoting and limited time exercises in light of high data stockpiling and encoding limit of QR codes.

In the mid 2000s, QR codes immediately acquired excellent standing in the Europe, and in North America before long, with advertisers and makers putting QR codes on numerous items and administrations from names of wine to cleanser jugs to candy bars.5 In China, QR codes are utilized in numerous day by day exercises including promoting, social stage (like WeChat), m-installment, internet shopping, and ticket the board.

PRIOR ART SEARCH

Numerous advertisers have been continually bantering with regards to whether or not QR codes are advantageous to their showcasing endeavors. QR codes are speedy and simple medium to take shoppers from the physical to the virtual world, that is, carrying on the web content to a versatile crowd in only seconds.

The review through perception of past and existing QR codes showcasing efforts of organizations, interviews with the two advertisers and purchasers and audit of different reports and articles uncovered six fundamental variables alluded to as the "Five W's and H" that should be all around thought to be preceding endeavor a promoting effort utilizing QR codes.

For a QR code mission to be effective, the review proposes an organization needs to have clear goals, know who they are focusing on, give significant substance, place code at suitable areas for simplicity of examining, settle on the perfect opportunity for the mission lastly execute it in an imaginative manner to stand apart from the group.

The point of this review is to assess the effect of speedy reaction (QR) codes on consumer loyalty and buy expectation in the internet shopping setting based on apparent stream. This concentrate likewise showed the effect of QR codes on apparent stream, trailed by apparent stream on buy aim and consumer loyalty, lastly the mix effect of seen stream and consumer loyalty on buy expectation. Boost Organism–Response hypothetical model was utilized in this examination.

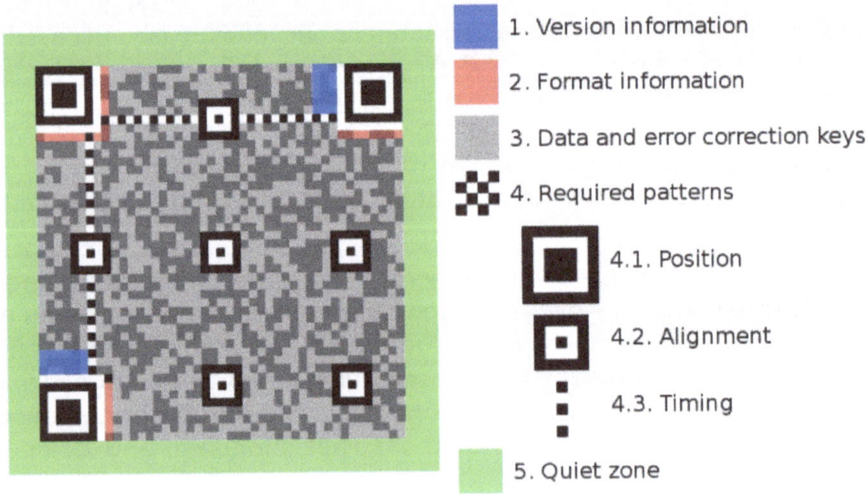

Fig. 3: QR-Code Notification

Information were gathered by online surveys from 420 legitimate respondents who bought an item online through QR code. The covariance-based underlying condition demonstrating approach was utilized to investigate the primary model and estimations.

This review showed that QR codes enormously affect buy expectation and consumer loyalty. The discoveries of this article affirmed that QR codes impact apparent stream, joined impact of them, thus, affected internet based customers' fulfillment and, at last, buy expectation.

These discoveries are essentially huge for the two advertisers and clients. Advertisers should utilize QR codes as an inserted publicizing device with specific URLs that will prompt greater productivity for shopping office. Other than clients can likewise get familiar with their conduct in regards to QR codes in the web based shopping setting. This exploration has incredible genuine ramifications in the field of Internet shopping, web based showcasing, retailing, advertising advancement, and shopper conduct.

Inserted QR codes with publicizing are exceptionally useful for promoters to share and impact consumer loyalty and buy expectation. Sharing data by means of QR codes will impact clients to draw in and share data with different clients in the internet based local area, which will assist with expanding the volume of deals quickly. Supposedly there is no exploration to assess the genuine effect of QR codes on buy expectation and consumer loyalty based on apparent.

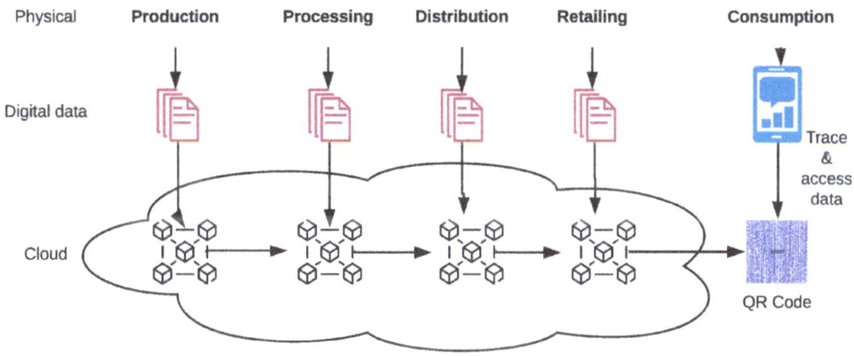

Fig. 4: QR-Code Notification Based Cloud Marketing.

OBJECTIVES

The objective of the invention is to provide a "|Customer| Preferences on QR-Code Notification Based Digital Marketing" is a || that is not the source or originator of a |QR code| is empowered to give a shopper a determination of related substitute substance instead of satisfying recognized in the |QR code.|

The other objective of the invention is to provide a strategy is |carried| out on a PC |framework| to get extra substance data addressed by a QR code removed by a buyer's |QR peruse| gadget from a |media| show containing the |QR code| and the QR code is |produced| by a wellspring of the media. An area of substitute substance from a hostage climate connected with the extra substance data is recognized.

The other objective of the invention is to provide aarea of the substitute substance might be |recognized| dependent on the decoded data from the QR code. The area of the |substitute| substance is sent to the QR peruse gadget. The point of this invention is to decide client inclinations towards various procedures of Digital Marketing used to advance offer of electronic gadgets.

The other objective of the invention is to provide aAdvanced Marketing apparatuses offer wide scope of chances for organizations to draw in and hold buyers |however| they need to know how their clients assess their advertising endeavor. To analyze which promoting procedures buyers like in this invention we utilize Choice Based Conjoint Analysis (CBC).

The other objective of the invention is to provide a strategy requests the respondents to pick between various mixes from promoting components, which addresses common decision-making issue they face in day-to-day existence. |Likewise| |CBC| offers us the chance to examine possible connections between given components.

The other objective of the invention is to provide a Knowing, which blend is-viewed as the most favoured one, could assist organizations with distributing their assets and make promoting procedure the s that will guarantee constructive outcomes on their business execution.

SUMMARY

Determined to foster images with extended limit and diminished size, two-layered codes were | |designed as a progression of the normal one-layered standardized tags. Various methodologies can be arranged into composite codes, stacked codes, spot codes and lattice codes. QR codes are a variety of network codes. Further progressions are 3D codes with the extra aspect shading profundity and 4D codes with multi-successive showcase.

Picking a specific code type relies upon the necessities of the application situation. This invention will zero in on QR codes, because of their specialized benefits (for example consolidating positive qualities of a few code types) and level of normalization. QR Codes are like scanner tags utilized by retailers to follow stock and value items at the retail location.

The huge contrast between QR codes and standardized tags is how much information QR codes can hold and share. QR Code contains data in both the vertical and even headings, though a standardized tag contains information in one bearing as it were. QR Code holds an extensively more prominent volume of data than a standardized tag. It is alluded to as QR codes due to its capacity to permit substance to be decoded at a high velocity.

QR Code is a sort of 2-D (two-layered) symbology created by Denso Wave and delivered in 1994 with the essential point of being an image that is effectively deciphered by scanner gear. Copyrights of QR codes are deferred by its patent holder, Denso Wave, which implies they can be uninhibitedly made and utilized for both individual and business use.

QR code Marketing Campaigns – Case Studies There are different brands/organizations that have utilized QR code in exceptionally essential ways and receiving assorted rewards. The following are a portion of these advantages: I. Expanded brand mindfulness II.

Increment brand support through verbal (WOM) / online connection buzz creation III. Increment client commitment IV. Construct tremendous client information base V. Increment deals VI. Client maintenance VII. Less advertising cost.

QR codes are persistently filling in fame and continually being utilized in more imaginative and intriguing ways. Advertisers are putting QR codes on nearly everything – bulletins, bundles, print promotions, on racks, transport covers, film tickets, magazines, on shirts, business cards, on skin (tattoo), in exhibition halls, libraries, cafés, and so forth Many are of the idea to put QR code in numerous areas as conceivable to create more filtering of the QR code. Many organizations are setting their QR codes at exceptionally fitting and invaluable spots whiles others are failing to understand the situation.

The item or administration included assumes a crucial part in where you need to put your QR code. Best Buy places QR codes on their sticker prices in their stores where purchasers can filter them and it will take them to a determination page on the web. McDonald's coverings have QR codes on them, where the code can be filtered for fixings or nourishing substance.

The McDonalds model shows how the arrangement is truly significant. Envision assuming McDonalds should place such a QR code on a bulletin some place in a recreational area, what great or worth will it bring to somebody who is in the recreation area at that specific time. Preferably the QR code ought to be set in a spot that supplements the activity set off by the QR code.

The code ought to likewise be in areas that will be not difficult to examine. I ran over QR codes put on moving targets and parkway announcements and was considering the way in which such a code can be examined.

A troublesome spot to output will deter shoppers from filtering your QR code. It ought to likewise be in areas with web access inclusion. Significantly, the code ought to be place in an area for customers to pause for a minute to filter.

Tweaked publicizing through PURLs with QR codes can assist with trying not to send unseemly messages to explicit clients or gathering of clients and impact buyers to buy the offered products/administrations. Additionally PURLs can assist with further developing shoppers' positive inclination in regards to new item and administration. Customized elevating likewise permits to acquire data for understanding future promoting and showcasing efforts.

QR codes permit buyers to visit the site whenever from anyplace to get to organizations' sites and the substance. Client can undoubtedly include with the site by visiting, loving, and recording bit of feedback; this sort of association is called clients' contribution in business which is vital for leading any business.

So by implication, QR codes help to develop business quickly. Development of data sharing stages and intelligent correspondence advances help to present new mechanical cycle in various ways.

QR code is one innovation that can be utilized as an intuitive showcasing medium by advertisers to publicize and advance their items or administrations. In addition, different benefits incorporate free utilization, fast checking capacity, and can be utilized by anybody without permit. So advertisers as intuitive publicizing and advancement media can pick QR codes.

Because of higher supporting abilities, those codes are thinking about as solid highlighting apparatuses for advertiser and shopping devices for purchasers. As cell phone advancements and its clients are expanding, QR codes are-utilized for an assortment of displaying purposes and versatile advertising is one of them.

DIAGRAM

Fig. 5: QR-Code Notification Based Digital Marketing.

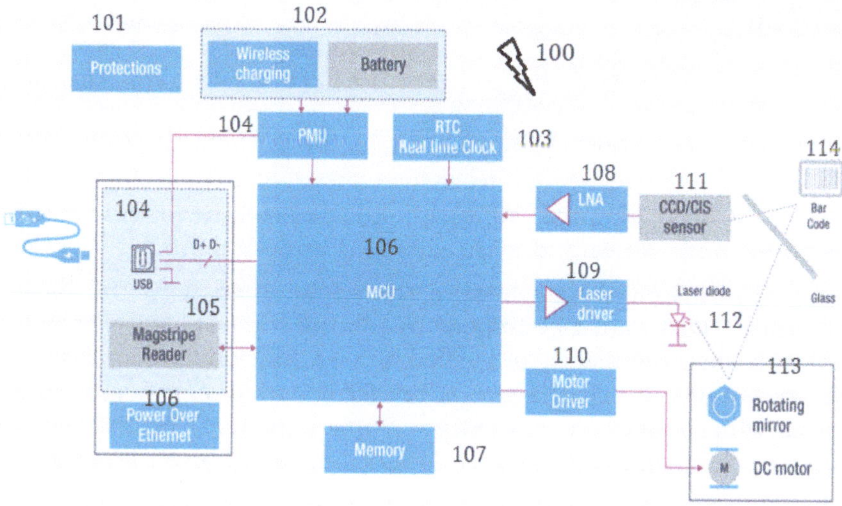

Fig. 6: QR-Code Notification Based Digital Marketing, Block Digram.

Fig. 7: QR-Code Notification Based Digital Marketing.

DESCRIPTION

Three distinct methodologies were -utilized to reveal these essential variables. Past examinations and articles (auxiliary information) done on QR codes from many sources like scholarly diaries, online distributions, sites, and so on were first surveyed to give the foundation data expected to profoundly comprehend the broadness and profundity of QR codes applications, advantages and impediments.

After this initial step, we were then furnished with the essential data and the right inquiries to pose to acquire top to bottom knowledge. Interviews were held with advertisers who have before or as of now embraced a QR code displaying effort. This was a significant basis for choosing who to meet for this review. Some were done on the telephone because of their tight timetable to set up an eye-to-eye meeting.

Their insight and conclusions served to obviously comprehend their impression of QR codes. Altogether, did ten top to bottom meetings; seven eye to eye and three phone interviews. At long last, to have a total comprehension of QR codes according to all viewpoints, irregular determination of customers were likewise asked a couple of open end inquiries to comprehend their mindfulness and impression of QR codes.

Purchasers included both male and female, went from various age gatherings (18 and over), business status and ethnicities. Illustration of inquiries "posed were| . Have you known about QR codes? Have you checked QR codes previously? Where was the QR code found- For what reason did you check the QR codes? What do you hope to get subsequent to checking? They were toward the end requested any extra data or assessment on QR codes they might want to share.

QR Codes and Perceived Flow

For the QR code to hold a larger number of characters of data than a one-layered conventional standardized identification, it can hold more item depiction related data including URL. QR code can impact client to visit advertisers' site. By visiting the site, client can get numerous item related data from advertisers' site. Introduced data on the site fundamentally affects buy aim of clients. Rosario Raymundo showed that the QR codes affect stream insight of buyer, which can again impact their dependability.

Framework quality, for example, an installed site in the QR code, ought to be improved because helpless framework quality cannot affect buy expectation and fulfillment of buyer. Value alludes to the shoppers' own interior assessment with respect to the need and need satisfaction limit of explicit item or administration. Attainability is the level of common sense association with the article or circumstance and agreeableness still up in the air with the level of need fulfillment ability.

QR code's helpfulness, possibility, and worthiness can be utilized as data convey devices appropriate to the clinical circumstance in a "thorough, ideal, student focused way, and cutting-edge." Prior examinations have tracked down that site's ease of use, usefulness, and adequacy have positive effect on apparent stream.

Subsequently, this review accepts that positive impression of QR codes conveyed to their stream experience empower them to engage themselves in their internet shopping exercises. To give stream insight to purchaser by QR with credits (handiness, plausibility, and adequacy), QR codes should be conveyed or shared by the advertiser in the internet shopping setting and so forth.

Research Outcome

Our Invention "|Customer| Preferences on QR-Code Notification Based Digital Marketing" is a || that is not the source or originator of a |QR code| is empowered to give a shopper a determination of related substitute substance instead of satisfying recognized in the |QR code.| A strategy is |carried| out on a PC |framework| to get extra substance data addressed by a QR code removed by a buyer's |QR peruse| gadget from a |media| show containing the |QR code| and also the QR code is |produced| by a wellspring of the media. An area of substitute substance from a hostage climate connected with the extra substance data is recognized. The area of the substitute substance might be |recognized| dependent on the decoded data from the QR code. The area of the |substitute| substance is sent to the QR peruse gadget. The point of this invention is to decide client inclinations towards various procedures of Digital Marketing used to advance offer of electronic gadgets. The interest in this sort of items is |relied upon| to increment later on, thusly it is fundamental for organizations to track down the most effective way to stand apart from the opposition |Some| |what| recently, and the advancement of current innovation has made Digital Marketing the main method of association among organizations and their clients. Advanced Marketing apparatuses offer wide scope of chances for organizations to draw in and hold buyers |however| they need to know how their clients assess their advertising endeavor. To analyze which promoting procedures buyers like in this invention we utilize Choice Based Conjoint Analysis (CBC). This strategy requests the respondents to pick between various mixes from promoting components, which addresses common decision-making issue they face in day-to-day existence. |Likewise| |CBC| offers us the chance to examine possible connections between given components. Knowing which blend is-viewed as the most favoured one could assist organizations with distributing their assets and make promoting procedure the s that will guarantee constructive outcomes on their business execution

- According to claim1# the invention is to a " |Customer| Preferences on QR-Code Notification Based Digital Marketing" is a || that is not the source or originator of a |QR code| is empowered to give a shopper a determination of related substitute substance instead of satisfying recognized in the |QR code.|
- According to claim1,2,3# the invention is to a strategy is |carried| out on a PC |framework| to get extra substance data addressed by a QR code removed by a buyer's |QR peruse| gadget from a |media| show containing the |QR code| and also the QR code is |produced| by a wellspring of the media. An area of substitute substance from a hostage climate connected with the extra substance data is recognized.
- According to claim1,2,3# the invention is to a area of the substitute substance might be |recognized| dependent on the decoded data from the QR code. The area of the |substitute| substance is sent to the QR peruse gadget. The point of this invention is to decide client inclinations towards various procedures of Digital Marketing used to advance offer of electronic gadgets.
- According to claim1,2,3,5# the invention is to interest in this sort of items is |relied upon| to increment later on, thusly it is fundamental for organizations to track down the most effective way to stand apart from the opposition |Some| |what| recently, and the advancement of current innovation has made Digital Marketing the main method of association among organizations and their clients.
- According to claim1,2,4,5 #, the invention is to a Advanced Marketing apparatuses offer wide scope of chances for organizations to draw in and hold buyers |however| they need to know how their clients assess their advertising endeavor. To analyze which promoting procedures buyers like in this invention we utilize Choice Based Conjoint Analysis (CBC).
- According to claim1,2,4,6 #, the invention is to a strategy requests the respondents to pick between various mixes from

promoting components, which addresses common decision-making issue they face in day-to-day existence. |Likewise| |CBC| offers us the chance to examine possible connections between given components.

- According to claim1,2,3,4,5# the invention is to a Knowing which blend is-viewed as the most favored one could assist organizations with distributing their assets and make promoting procedure the s that will guarantee constructive outcomes on their business execution.

REFERENCES

[1] R. Focardi, F. L. Luccio, and H. A. M. Wahsheh, "Usable security for QR code," Journal of Information Security and Applications, vol. 48, p. 102369, 1 October 2022.

[2] Y. Wang, C. Sun, P. Kuan, C. Lu, and H. Wang, "Secured graphic QR code with infrared watermark," in 2018 IEEE International Conference on Applied System Invention (ICASI), pp. 690-693, 2019

[3] Y. Cheng, Z. Fu, and B. Yu, "Improved Visual Secret Sharing Scheme for QR Code Applications," IEEE Transactions on Information Forensics and Security, vol. 13, no. 9, pp. 2393-2403, 2020.

[4] Z. Fu, Y. Cheng, and B. Yu, "Visual Cryptography Scheme with Meaningful Shares Based on QR Codes," IEEE Access, vol. 6, pp. 59567-59574, 2018.

[5] J. Song, K. Gao, X. Shen, X. Qi, R. Liu, and K.-K. R. Choo, "QRFence: A flexible and scalable QR link security detection framework for Android devices," Future Generation Computer Systems, vol. 88, pp. 663-674, 2018.

[6] Y. Qin, Z. Wang, H. Wang, and Q. Gong, "Binary image encryption in a joint transform correlator scheme by aid of run-length encoding and QR code," Optics & Laser Technology, vol. 103, pp. 93-98, 2018.

[7] Y. Wei, A. Yan, J. Dong, Z. Hu, and J. Zhang, "Optical image encryption using QR code and multilevel fingerprints in gyrator transform domains," Optics Communications, vol. 403, pp. 62-67, 2019.

[8] P. Nazemzadeh, D. Fontanelli, D. Macii, and L. Palopoli, "Indoor Localization of Mobile Robots Through QR Code Detection and Dead Reckoning Data Fusion," IEEE/ASME Transactions on Mechatronics, vol. 22, no. 6, pp. 2588-2599, 2020.

[9] S. Demir, R. Kaynak, and K. A. Demir, "Usage Level and Future Intent of Use of Quick Response (QR) Codes for Mobile Marketing among College Students in Turkey," Procedia - Social and Behavioral Sciences, vol. 181, pp. 405-413, 2015

CITATIONS

- Jain, A., Dwivedi, R. K., Alshazly, H., Kumar, A., Bourouis, S., & Kaur, M. (2022). Design and Simulation of Ring Network-on-Chip for Different Configured Nodes. CMC-COMPUTERS MATERIALS & CONTINUA, 71(2), 4085-4100.
- Jain, A., Kumar, A. and Sharma, S., 2015. Comparative Design and Analysis of Mesh, Torus and Ring NoC. Procedia Computer Science, 48, pp.330-337.
- Jain, A., Dwivedi, R., Kumar, A. and Sharma, S., 2017. Scalable design and synthesis of 3D mesh network on chip. In Proceeding of International Conference on Intelligent Communication, Control and Devices (pp. 661-666). Springer, Singapore.
- Jain, A., Gahlot, A.K., Dwivedi, R., Kumar, A. and Sharma, S.K., 2018. Fat Tree NoC Design and Synthesis. In Intelligent Communication, Control and Devices (pp. 1749-1756). Springer, Singapore.
- Jain, A., Dwivedi, R., Kumar, A. and Sharma, S., 2017. Scalable design and synthesis of 3D mesh network on chip. In Proceeding of International Conference on Intelligent Communication, Control and Devices (pp. 661-666). Springer, Singapore.

CHAPTER - 08

E-Authentication using IoT

Dr. Lokesh Kumar
Assistant Professor, Tula's Institute, Dehradun, Uttarakhand
Ms. Monika Belwal
Assistant Professor, Tula's Institute, Dehradun, Uttarakhand
Mr. Anurag Kumar
Assistant Professor, Tula's Institute, Dehradun, Uttarakhand

ABSTRACT

The new worldview Internet of Things interfaces all items to the Internet and permits them to detect cum speak with themselves and the outside world. This environment upgrades efficiency, makes new plans of action, and creates new income streams. It empowers a gigantic trade of information which was never accessible however posts difficulties in acquiring clients' data a solid way. Defending the associated IoT gadgets and organizations are much essential and truly challenging due to the idea of IoT framework. This paper audits the generally existing IoT verification writings for secure transmission of information and presents a logical study of the current work. And furthermore, the creators frame the expected issues and difficulties of verification in IoT for additional exploration. This paper examines the different verification frameworks carried out for upgraded security and private reposition of a person's login qualifications. The initial segment of the paper depicts the multifaceted verification (MFA) frameworks, which, however not appropriate to the field of Internet of Things, gives extraordinary security to a client's qualifications. MFA is trailed by a short portrayal of the functioning instrument of association of outsider clients with private assets over the OAuth convention structure and an investigation of the appointment based validation framework in IP-based IoT.

Keywords: *Authentication; IoT Environment; Analytical Survey; Issues and challenges in IoT Authentication.*

INTRODUCTION

In the start of this time, living being interface as well as gadgets speak with one another. This sort of gadget correspondence is called Internet of things (IoT) and has entranced the consideration as acknowledged as the future world. In an IoT climate, more gadgets are associated step by step. This development carries a few advantages to complete everyday errands. Yet, these advantages become a gamble, as the programmers and digital crooks are to an ever increasing extent.

These gigantic security dangers stand out enough to be noticed from the scientists and academicians. Giving a right security to the Internet of Things will assemble trust in the inexorably associated world. Along these lines, this exploration thinks about validation of IoT climate as its center and chips away at planning a lightweight confirmation component for IoT gadgets and clients.

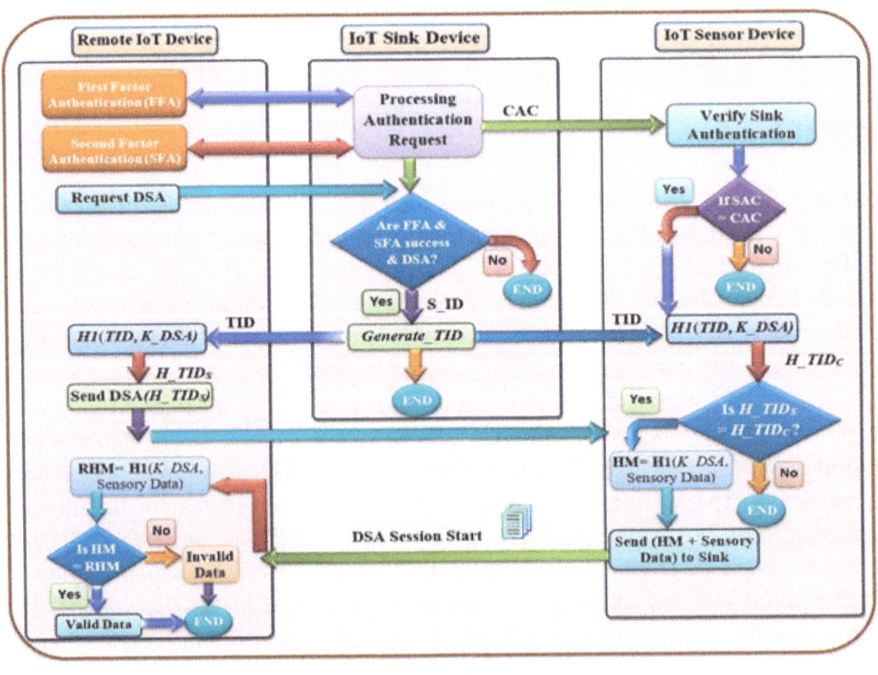

Fig. 1: E-Authentication using IoT

Wide reception of consistent combination of informal organization with any electronic gadgets like an advanced mobile phone, wellness tracker and so on has worked with the cooperation of staff with anybody over the world with simply the snap of a button. This has made ready for sharing of individual wellbeing related information [1], online social record and a lot more with anybody in any mainland and has opened the way for social cooperation and contest over the Internet.

However the correspondence occurring between the shrewd items and the web-based entertainment is helpless to capture from different outsider gatecrashers, which might prompt loss of protection and openness of validation subtleties to undesirable staff. Subsequently an improved and more vigorous validation framework from logging inside one's internet based record will guarantee a lot more secure perusing experience and online trade of data.

AUTHENTICATION OF IOT DATA

Validation is the method involved with perceiving clients and gadgets in an organization and restricting induction to approved people and non-controlled gadgets. This cycle really depends on username and secret key and don't work with unattended gadgets. Confirmation can be of one-way verification and shared validation.

In an IoT climate, the item confirms the server as well as the other way around. Here, the server is overseeing security declarations given by the IoT gadgets. Along these lines, just real clients and servers can partake in the data move. portrays the correspondence situation of IoT climate.

Encryption calculations should be less energy consuming however more effective, and productive key circulation plans should be clear cut [3][1]] Most security plans for IoT gadgets utilize public key cryptography. In the event that the key is hacked, the security of the whole framework will be impacted [1]. Secure and dependable association of asset obliged gadgets are troublesome, as a result of the heterogeneity in IoT networks [2].

Authentication is vital with regards to IoT. In any case, the confirmation conspire should answer the qualities of the Internet of Things [6]. Major prerequisites of IoT is the security enhancements and protection [1][8]. The impediment of organization assets made hard to gather information from various IoT gadgets progressively. In this manner, it is crucial for control the information assortment to work on the exhibition of the verification framework. [2][6]

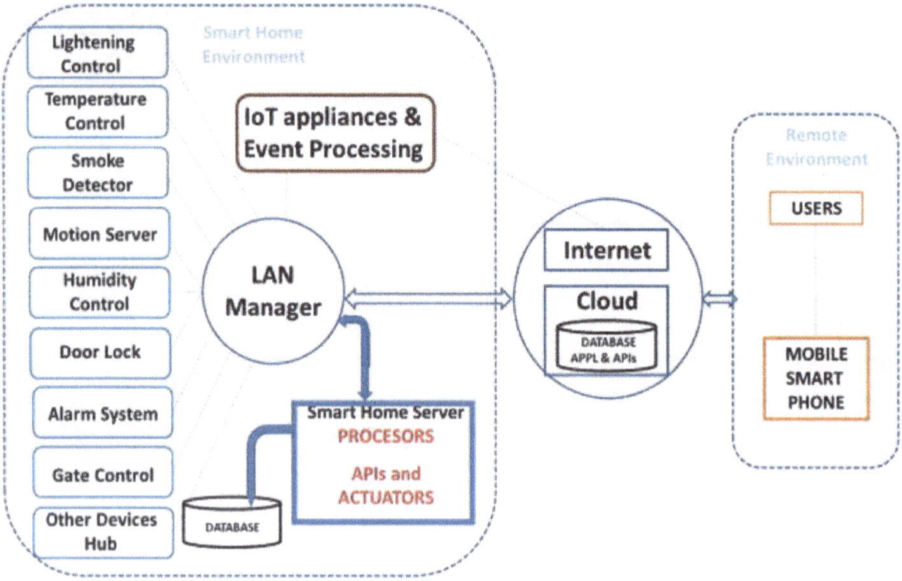

Fig. 2: E-Authentication using IoT, Process

The current correspondences in IoT has restricted inbuilt single-factor verification security system which isn't adequate to moderate the dangers and requires expanded validation conspire. Consequently, IoT engineering necessities to imagine a two-factor verification plan to meet fundamental security prerequisites, for example, classification, trustworthiness and accessibility of the gadgets and their correspondences to visualize shrewd applications [2][3].

Security and protection are significant areas of worry in IoT-based medical care applications as most gadgets and their interchanges are remote in nature. Giving hearty and secure information correspondence among medical services sensors, actuators, patients, and parental figures are urgent [2][1]IoT gadgets are compelled regarding battery duration, handling power, and memory and work in a remote climate.

It makes various systems administration challenges [1][4]Because of the asset obliged climate, exemplary security components can't be applied on IoT gadgets since, they consume more energy [28] ü Security is one of the significant issues that impede the reception of IoT Cloud suppliers and making of IoT gadgets entrusted with the Cloud framework is a major concern

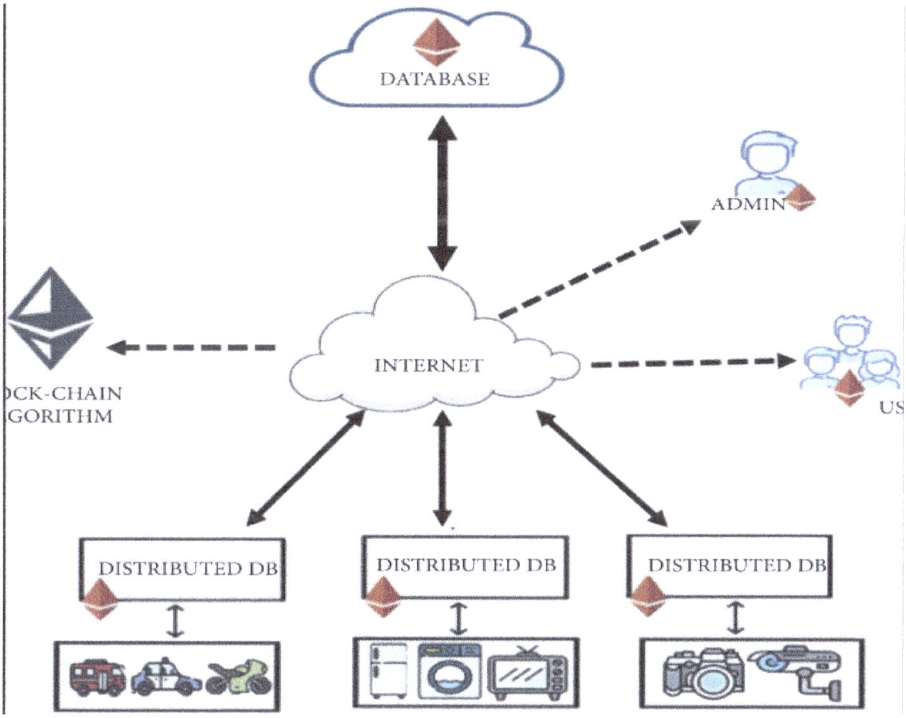

Fig. 3: E-Authentication using IoT, Method

AUTHENTICATION

In any informal community associated with a brilliant item, a connection between the web-based gateway and the Internet associated gadget is started by the transfer of the enrolled client ID and its relating secret word to a web-based server. Verification is the most common way of affirming one's character. Two-factor validation or 2FA is a method of login where a client is expected to give extra data to sign in than only the secret word. Utilizing just the secret phrase to login empowers vindictive assailants to have simple access into the framework as it addresses single snippet of data just [1][4].

In validation frameworks, all the transmission of information from a client's shrewd item to the web-based server can be presented to undesirable faculty through capture. Thusly, from a security stance, the most well-known confirmation frameworks neglect to ensure a safeguard strategy for getting the login data far from the hands of the general population for keeping up with protection and security for the client. With the rising utilization of cell phones by the buyers for banking, shopping [1][7] and so on the requirement for security concerns have arisen which thusly has made a premium in multifaceted validation.

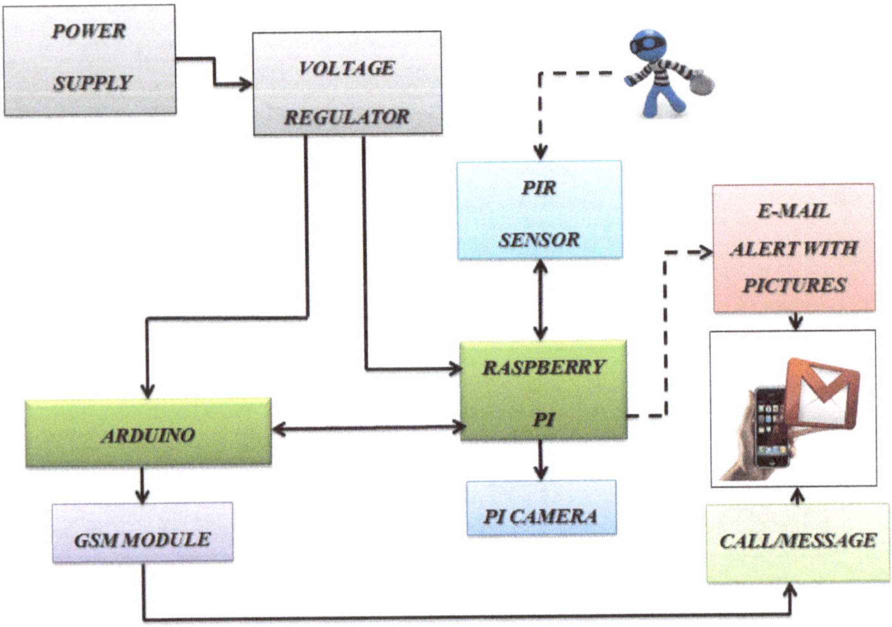

Fig. 4: E-Authentication using IoT, Structure

If there should be an occurrence of multifaceted validation, which requires more than one type of confirmation for check of authenticity, gives an extra layer of assurance against security break. Here separated from giving username and secret key by the client, an extra validation code is shipped off the client's cell phone.

TWO-STEP AUTHENTICATION IN NON-IOT DEVICES

In a portion of the main tech organizations offering on the web benefits, the organizations offer the decision of Multi-factor verification framework for improved security. In this sort of confirmation framework, check of explicit blend of various parts happens prior to validating a person. On the off chance that one of the parts is missing/wrong, the personality of the individual isn't acknowledged as enrolled in the framework.

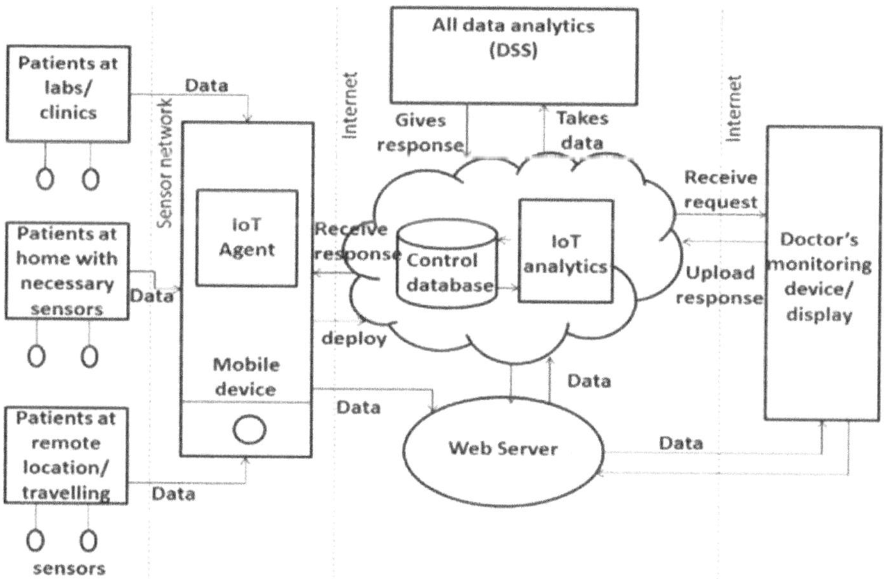

Fig. 5: E-Authentication using IoT, Block Working

For improving the protection from interlopers, while moving information, information concealing strategy [8] can be carried out by the organizations to forestall information robbery. One of the most conspicuous multifaceted confirmation frameworks is the cell phone two-factor validation [2]. In this framework, a client gets to his record by contributing the enrolled data alongside a unique one-time substantial secret word [5] comprised of blend of digits.

The last secret word is shipped off the client through email, SMS or some other explicit application intended to work with the acquirement of the OTP (One Time Password) from the host server. To guarantee the security of the OTP, its legitimacy is restricted to an extremely brief length of time [7] past which the OTP terminates and the client needs to return the individual data comparing to which another OTP is mentioned, created and shipped off the client from the server

This is a verification technique for Internet of Things proposed by Hummen et al. [3][1]. DTLS represents Datagram Transport Layer Security. In an assigned DTLS association foundation, a designation server is utilized for isolating the underlying foundation of association from the resulting application information assurance. Here the designation server makes the arrangement for an obliged gadget with every one of the important settings of safety for got correspondence.

These designation servers contain no pre-shared secret keys for the communicators yet lays out security settings on-request. Initial an expert key in engraved in the compelled gadget by the assignment server for transmission of safety settings forward and backward between the gadgets.

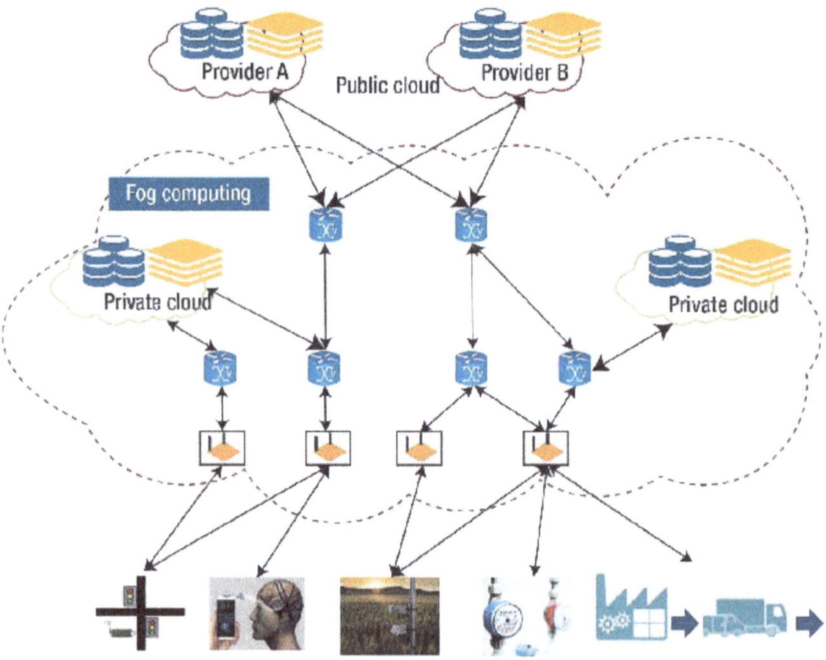

Fig. 6: E-Authentication using IoT, Block Working

This engraving is done when the assignment server is bootstrapped to the nearby organization space. Under the presumption of organization by a typical administrator, the between association between the compelled gadget and a remote end-point is laid out by the guidance of the administrator to the designation server to lay out a DTLS association with the previously mentioned end-point.

CONCLUSIONS

This paper presents the outline of confirmation in IoT climate and its exploration challenges. A wide assortment of literary works were introduced. Each study was audited to figure out the issues and issues in the security of IoT conditions. As indicated by the above introduced writing study, it is recognized that security in IoT is a significant issue when it turns into a reality.

Thus, IoT security framework should be intended for improving the verification and approval to convey better security administration. In the event that the confirmation component is more grounded and deep rooted, it will forestall numerous security dangers like snoopping, pantomime and replay assault, and so forth. In addition, the verification systems ought to be quick and light weighted without compromising security.

REFERENCES

[1] Anjali Yeole, Sadaf Ahmedi, KirtiMadhwani, SnehaSahijwani, Pooja Talreja, "A Robust Scheme for Secure Communication in Internet of Things", International Journal of Innovative Research in Computer and Communication Engineering, Vol. 3, Issue 11, November 2022, pp. 10401 – 10406

[2] Ajit A. Chavan and Mininath K. Nighot, "Secure and Cost-effective Application Layer Protocol with Authentication Interoperability for IOT", Science Direct, Elsevier, Procedia Computer Science, Vol, 78, 2020, pp. 646 – 651.

[3] ByungMun Lee, "Authorization Protocol using a NFC P2P mode between IoT device and Mobile phone", Advanced Science and Technology Letters, Vol.94, 2015, pp.85-88, http://dx.doi.org/ 10.14257/astl. 15.94.18

[4] Debiao He and SheraliZeadally, "An Analysis of RFID Authentication Schemes for Internet of Things in Healthcare Environment Using Elliptic Curve Cryptography", IEEE Internet of Things Journal, Vol. 2, No. 1, February 2019, pp. 72 -83.

[5] C Gehrmann, Marco Tiloca, and RikardHoglund, "SMACK: Short Message Authentication ChecK Against Battery Exhaustion in the Internet of Things", 12th International Conference on Sensing, Communication, and Networking (SECON), IEEE, 2019, pp. 274 – 282.

[6] Hamza Khemissa and DjamelTandjaoui, "A Lightweight Authentication Scheme for E-health applications in the context of Internet of Things", IEEE Digital Library, 2015, pp. 90 -95, DOI: 10.1109/NGMAST.2018.31

[7] Hamza Khemissa, DjamelTandjaoui, "A Novel Lightweight Authentication Scheme for heterogeneous Wireless Sensor Networks in the context of Internet of Things", Wireless Telecommunications Symposium (WTS), IEEE, 2016, DOI: 978-1-5090-0314-3, pp. 1-6.

CITATIONS

- Jain, A., Kumar, A. and Sharma, S., 2015. Comparative Design and Analysis of Mesh, Torus and Ring NoC. Procedia Computer Science, 48, pp.330-337.
- Jain, A., Dwivedi, R., Kumar, A. and Sharma, S., 2017. Scalable design and synthesis of 3D mesh network on chip. In Proceeding of International Conference on Intelligent Communication, Control and Devices (pp. 661-666). Springer, Singapore.
- Jain, A., Gahlot, A.K., Dwivedi, R., Kumar, A. and Sharma, S.K., 2018. Fat Tree NoC Design and Synthesis. In Intelligent Communication, Control and Devices (pp. 1749-1756). Springer, Singapore.
- Jain, A., Dwivedi, R., Kumar, A. and Sharma, S., 2017. Scalable design and synthesis of 3D mesh network on chip. In Proceeding of International Conference on Intelligent Communication, Control and Devices (pp. 661-666). Springer, Singapore.
- Sharma, S.K., Jain, A., Gupta, K., Prasad, D. and Singh, V., 2019. An internal schematic view and simulation of major diagonal mesh network-on-chip. Journal of Computational and Theoretical Nanoscience, 16(10), pp.4412-4417.

CHAPTER - 09

E-Authentication using IoT and ML

Madhuri Babar

Assistant Professor, JD College of Engineering & Research, Nagpur MH, India
madhubabar500@gmail.com

Dr. Neeta N Thune

Assistant Professor, Zeal College of Engineering & Research, Pune, MH, India.
neeta.ghune@zealeducation.com

Madhuri Sonkhaskar

Asst. Professor, ZCOER, madhurisonkhaskar@gmail.com

ABSTRACT

With the continuous endeavors for far reaching Internet of Things (IoT) reception, one of the key variables upsetting the wide acknowledgment of IoT is security. Getting IoT organizations, for example, the electric power network or water supply frameworks has arisen as a significant public and worldwide need. To address the security issue of IoT, a few investigations are being completed that include the utilization of, however are not restricted to, blockchain, man-made consciousness, and edge/haze registering. Confirmation and approval are essential parts of the CIA group of three to safeguard the organization from malignant gatherings. Notwithstanding, existing approval and validation plans are not adequate for dealing with security, because of the size of the IoT organizations and the asset obliged nature of gadgets. To defeat difficulties because of different requirements of IoT organizations, there is a huge interest in involving AI methods to aid the validation and approval process for IoT. In this paper, ongoing advances in validation and approval methods for IoT networks are explored. In light of the survey, we present a scientific categorization of verification and approval plans in IoT zeroing in on AI based plans. Utilizing the introduced scientific categorization, an intensive investigation is given of the verification and approval (AA) security dangers and difficulties for IoT. Besides, different standards to accomplish a serious level of AA flexibility in IoT executions to upgrade IoT security are assessed. Finally, a definite conversation on open issues, difficulties, and future examination headings is introduced for empowering secure correspondence among IoT hubs

Key: *E-Authentication,IoT, ML, Investigation ,Verification, Utilizing, Introduced*

BACKGROUND

The future Internet of Things (IoT) will have a significant moderate, business and social impact on our lives. The taking an interest center in IoT networks are for the most part resource constrained, which makes them drawing centers for advanced attacks. In such way, expansive undertakings have been made to address the security and insurance issues in IoT networks fundamentally through regular cryptographic procedures.

Anyway, the uncommon meal acteristics of IoT center points render the ongoing game plans lacking to integrate the entire security scope of the IoT associations. This is, essentially somewhat, considering the resource objectives, heterogeneity, huge consistent data created by the IoT devices, and the extensively novel direct of the associations.

As such, Machine Learning (ML) and Deep Learning (DL) strategies, which can give embedded understanding in the IoT devices and associations, are used to adjust to different security issues. In this paper, we purposefully review the security necessities, attack vectors, and the ongoing security deals with the IoT associations. We then, shed light on the openings in these security courses of action that call for ML and DL moves close.

We furthermore look at comprehensively the ongoing ML and DL deals with any consequences regarding watching out for different security issues in IoT associations. At long last, considering the distinct assessment of the ongoing plans in the composition, we discuss the future investigation direction for ML-and DL-based IoT security.

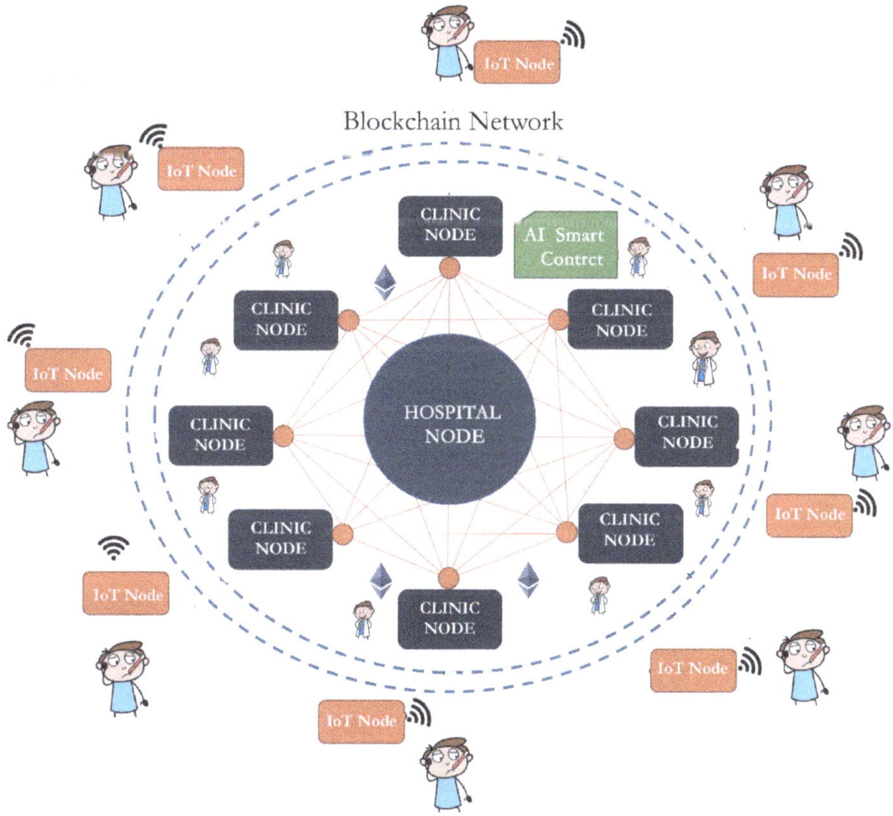

Fig. 1: E-Authentication using IoT and ML , Process

INTRODUCTION

Data made by the IoT devices is colossal and thusly, customary data variety, storing, and dealing with systems may not work at this scale. In addition, the sheer proportion of data can moreover be used for plans, practices, assumptions, and assessment. Additionally, the heterogeneity of the data created by IoT makes another front for the ongoing data dealing with frameworks.

Along these lines, to harness the value of the IoT-created data, new parts are required. In this novel situation, Machine Learning (ML) is seen as maybe the most suitable computational perspective to give embedded understanding in the IoT devices.

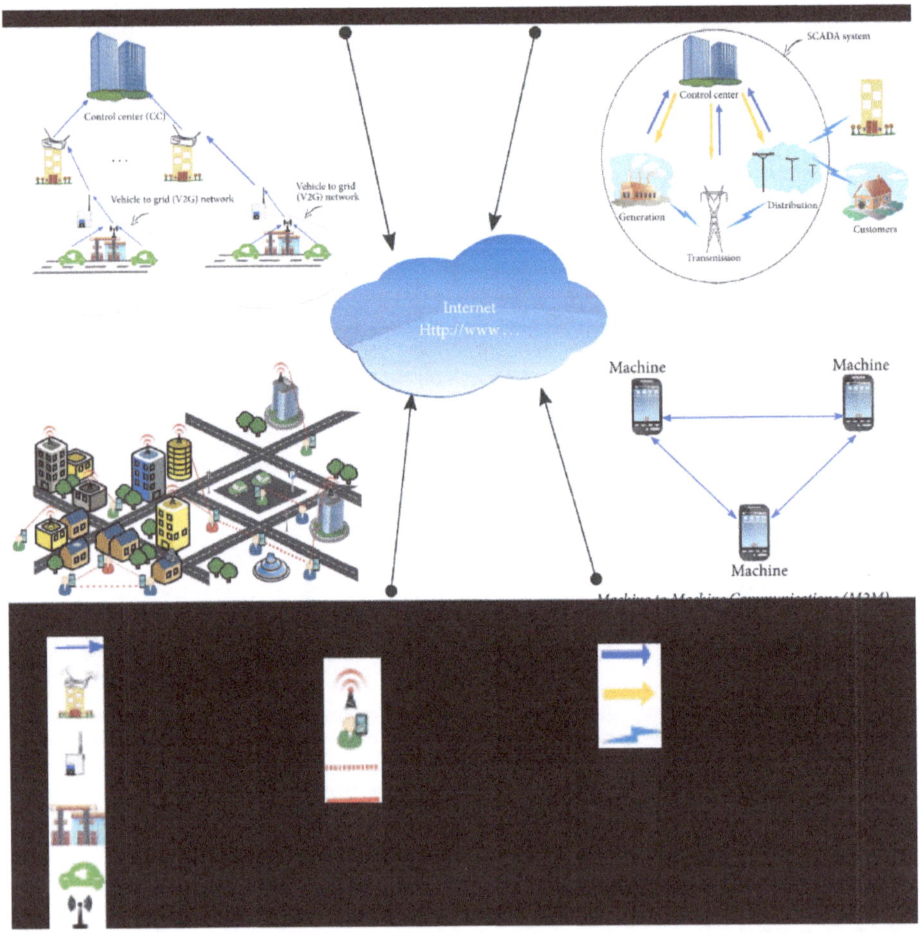

Fig. 2: E-Authentication using IoT and ML, Complete Structure

Correspondence in Nearness

One more striking component of IoT is the correspondence in closeness without including the central experts like base stations. Contraption toDevice correspondence (D2D) utilize the characteristics of feature point correspondence like Dedicated Short Range Communication (DSRC) and relative advances.

The plan of standard Internet is more arranged towards network-driven correspondence while actually the decou-pling of associations and organizations engage device driven as well as fulfilled driven correspondence which upgrades the IoT organization range.

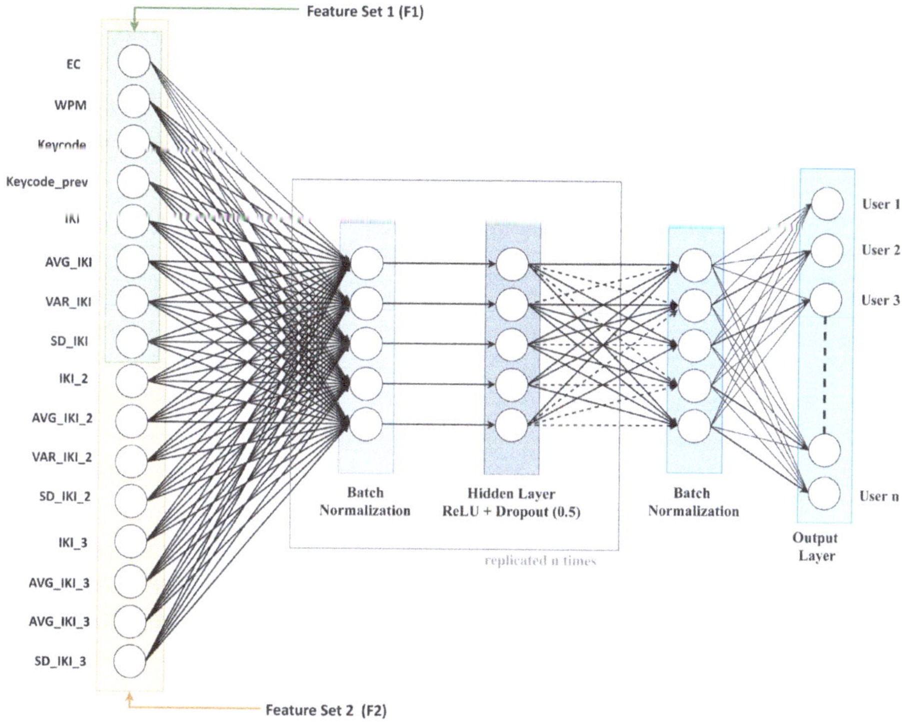

Fig. 3: E-Authentication using IoT and ML , Security

SECURITY CHALLENGES AND THREAT MODELS IN IOT

In a general sense, IoT uses an unprecedented method for managing good for vide buyers with different applications and organizations. The undeniable sending of tremendous number of contraptions also increase the attack surface. Of course, the way that IoT contraptions are (generally speaking) resource obliged; along these lines, it is absurd to expect to use complex security instruments against popular attacks.

Moreover, it merits zeroing in on that the principal Internet was not expected for IoT. Thusly, it is impera-tive to give IoT security on top of the ongoing security parts of the Internet and the essential advances. To this end, IoT uses different correspondence progresses, for instance, yet not confined to, IPv6, Zigbee, 6LoWPAN, Blue-tooth, Z-Wave, WiFi, and Near Field Communications (NFC), to give a few models .

Physical Attacks

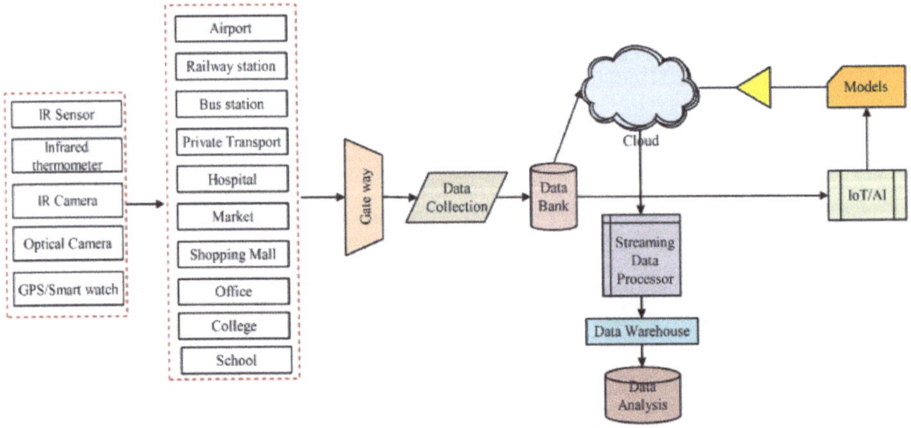

Fig. 4: E-Authentication using IoT and ML, Block Diagram

In genuine attacks, the aggressors have direct permission to the contraptions and control different pieces of the devices. To acquire induction to the real contraptions, social planning is potentially the most unquestionable methods where the attacker access the devices and perform certifiable attack that compasses from genuine damage to the device to tuning in, sidechannels, and other related attacks.

Despite how different advances are used at the genuine layer for IoT, the possibility of real attacks for the most part seem to be and require social planning like strategies. Moreover, to ship off real attacks, the aggressors ought to be in the closeness of the contraptions/gear with.

AUTHENTICATION AND ACCESS CONTROL IN IOT

This is one of the fundamental security essentials in IoT. The clients ought to be approved to use IoT applications or possibly benefits. Regularly, IoT applications and organizations rely upon data exchange across different stages. The data recuperated from the IoT contraptions is pre-taken care of, dealt with, and thereafter went through a decision genuinely steady organization to leave it.

These cycles could vary dependent upon the key IoT designing; nevertheless, data stream may be indistinct in these systems. Without loss of distortion, when an application or possibly a client needs a couple of data from an IoT contraption, the component (client or application) ought to be affirmed to IoT association and it ought to be guaranteed that the requester has required induction honors for the data. Differently, the sales to access such data will be denied.

CONCLUSION

IoT security and insurance are of crucial importance and expect a fundamental part in the commercialization of the IoT development. Regular security and insurance plans experience the evil impacts of different issues that are associated with the remarkable thought of the IoT associations. ML and even more unequivocally DL and DRL techniques can be used to engage the IoT devices to change in accordance with their dynamic environment.

These learning methodology can maintain self-assembling movement and moreover redesign the overall structure execution by acquiring and dealing with verifiable information from the environment (for instance human clients and IoT devices).

REFERENCES

[1] O. Novo, N. Beijar, and M. Ocak, "Capillary Networks - Bridging the Cellular and loT Worlds ," IEEE World Forum on Internet of Things (WF-IoT), vol. 1, pp. 571–578, December 2022

[2] F. Hussain, Internet of Things; Building Blocks and Business Modles. Springer, 2020.

[3] A. Al-Fuqaha, M. Guizani, M. Mohammadi, M. Aledhari, and M. Ayyash, "Internet of things: A survey on enabling technologies, protocols, and applications," IEEE Communications Surveys Tutorials, vol. 17, pp. 2347–2376, Fourthquarter 2019.

[4] J. Granjal, E. Monteiro, and J. S. Silva, "Security for the internet of things: A survey of existing protocols and open research issues," IEEE Communications Surveys Tutorials, vol. 17, pp. 12941312, thirdquarter 2019.

[5] A. Mosenia and N. K. Jha, "A comprehensive study of security of internet-of-things," IEEE Transactions on Emerging Topics in Com- puting, vol. 5, pp. 586–602, Oct 2018.

[6] J. Lin, W. Yu, N. Zhang, X. Yang, H. Zhang, and W. Zhao, "A survey on internet of things: Architecture, enabling technologies, security and privacy, and applications," IEEE Internet of Things Journal, vol. 4, pp. 1125–1142, Oct 2018.

[7] M. Ammar, G. Russello, and B. Crispo, "Internet of things: A survey on the security of iot frameworks," Journal of Information Security and Applications, vol. 38, pp. 8 – 27, 2018.

CITATIONS

- Ghai, D., Gianey, H.K., Jain, A. and Uppal, R.S., 2020. Quantum and dual-tree complex wavelet transform-based image watermarking. International Journal of Modern Physics B, 34(04), p.2050009.
- Jain, A. and Kumar, A., 2021. Desmogging of still smoggy images using a novel channel prior. Journal of Ambient Intelligence and Humanized Computing, 12(1), pp.1161-1177.
- Kumar, S., Jain, A., Shukla, A.P., Singh, S., Raja, R., Rani, S., Harshitha, G., AlZain, M.A. and Masud, M., 2021. A Comparative Analysis of Machine Learning Algorithms for Detection of Organic and Nonorganic Cotton Diseases. Mathematical Problems in Engineering, 2021.
- Misra, N.R., Kumar, S. and Jain, A., 2021, February. A Review on E-waste: Fostering the Need for Green Electronics. In 2021 International Conference on Computing, Communication, and Intelligent Systems (ICCCIS) (pp. 1032-1036). IEEE.
- Gupta, N., Jain, A., Vaisla, K.S., Kumar, A. and Kumar, R., 2021. Performance analysis of DSDV and OLSR wireless sensor network routing protocols using FPGA hardware and machine learning. Multimedia Tools and Applications, 80(14), pp.22301-22319.

CHAPTER - 10

Blockchain-based Voting Tracking using IoT

Dr. Geeta Arora

Head Department of Mathematics, DPG Degree College, Gurugram, Haryana
drgeetaarora81@gmail.com

Prof. (Dr.) Vivek Kumar

Vice Chancellor, Quantum University, Roorkee, UK, India

Prof (Dr.) B.K. Sarkar

MES's, Pillai HOC College of Engineering &Technology,
Rasayani Taluka Panvel, Dist, Navi Mumbai, Maharashtra 410207, India

ABSTRACT

Our Invention "Blockchain-based Voting Tracking using IoT" This invention work to use advantages of blockchain, for example, cryptographic establishments and straightforwardness to accomplish a powerful plan for e-voting. The invented plot adjusts to the necessities for e-casting a ballot conspires and accomplishes start to finish undeniable nature. The invention presents subtleties of the proposed e-casting a ballot conspire alongside its execution utilizing the Multichain stage. Web-based democracy is a pattern that is acquiring energy in present-day culture. It can diminish hierarchical expenses and increment citizen turnout. It dispenses with the need to print votingfrom inventions or open surveying stations—citizens can cast a ballot from any place there is an Internet association. Blockchain innovation came into the ground to beat these issues and offers decentralized hubs for electronic democratic and is utilized to create electronic democratic frameworks mostly due to their start to finish confirmation benefits. This innovation is a delightful substitution for customary electronic democratic arrangements with disseminated, non-renouncement, and security insurance attributes. The accompanying article gives an outline of electronic democratic frameworks dependent on blockchain innovation. As a result of this review, it was found that blockchain frameworks might assist with settling a portion of the issues that currently plague political decision frameworks. Then again, the frequently referenced issues in blockchain applications are security assurance and exchange speed. For a maintainable blockchain-based electronic democratic framework, the security of distant cooperation should be feasible, and for versatility, exchange speed should be tended to. Because of these worries, not set in stone that the current structures should be improved to be used in casting a ballot framework.

Key: *Blockchain, Voting, Tracking, IoT.*

FIELD

Our Invention is related to a Blockchain-based Voting Tracking using IoT.

BACKGROUND

Bitcoin stays the most recognized utilization of blockchain anyway specialists are quick to investigate the utilization of blockchain innovation to work with applications across various spaces utilizing advantages like non-disavowal, trustworthiness, and obscurity. In this invention, we investigate the utilization of blockchain to work with e-casting a ballot application with the capacity to guarantee citizen obscurity, vote honesty, and end-to-confirmation.

We accept e-casting a ballot can use from central blockchain highlights, for example, self-cryptographic approval structure among exchanges (through hashes) and public accessibility of circulated record of records. Blockchain innovation can assume a key part in the area of electronic democracy because of the intrinsic nature of protecting obscurity, keeping up with a decentralized and freely dispersed record of exchanges across every one of the hubs.

This makes blockchain innovation extremely proficient to manage the danger of using a democratic symbolic at least a couple of times and the endeavor to impact the straightforwardness of the outcome. The focal point of our exploration is to examine the central questions like citizen secrecy, vote privacy, and start to finish confirmation.

These difficulties structure the establishment of a proficient democratic framework safeguarding the trustworthiness of the democratic interaction. In this invention, we present our endeavors to investigate the utilization of blockchain innovation to look for answers for these difficulties. Specifically, our framework depends on the Prêt à Voter approach and utilizes an open-source blockchain stage, as the hidden innovation to foster our framework.

To ensure the namelessness and honesty of a vote, the framework creates a solid cryptographic hash for each vote exchange dependent on data explicit to an elector. This hash is additionally imparted to the elector utilizing encoded channels to work with a check.

Similar to the case with some other innovations, blockchain innovation has its downsides. In contrast to other conveyed arrangements, a blockchain is difficult to scale: An expanding number of hubs doesn't further develop network execution because, by definition, each hub needs to execute all exchanges, and this cycle isn't divided between the hubs.

In addition, expanding the number of validators impacts execution since it infers a more serious trade of messages during agreement. For a similar explanation, blockchain arrangements are defenseless against different disavowal of administration assaults.

Assuming a blockchain permits anybody to distribute brilliant agreements in an organization, the activity of the whole organization can be debilitated by basically placing a boundless circle in a shrewd agreement. An organization can likewise be assaulted by just sending a significant number of exchanges: sooner or later, the framework will get whatever else.

In digital money arrangements, all exchanges have an execution cost: the more assets an exchange uses, the costlier it will be, and there is an expense limit, with exchanges surpassing the edge being disposed of.

OBJECTIVES

The objective of the invention is to provide a "Blockchain-based Voting Tracking using IoT" This invention work to use advantages of blockchain, for example, cryptographic establishments and straight forwardness to accomplish a powerful plan for e-voting.

The other objective of the invention is to provide possibly diminish hierarchical expenses and increment citizen turnout. It dispenses with the need to print voting from inventions or open surveying stations—citizens can cast a ballot from any place there is an Internet association.

The other objective of the invention is to provide a Blockchain innovation that came into the ground to beat these issues and offers decentralized hubs for electronic democratic and is utilized to create electronic democratic frameworks mostly due to their start to finish confirmation benefits.

The other objective of the invention is to provide a result of this review, it was found that blockchain frameworks might assist with settling a portion of the issues that currently plague political decision frameworks. Then again, the frequently referenced issues in blockchain applications are security assurance and exchange speed.

The other objective of the invention is to provide a maintainable blockchain-based electronic democratic framework, the security of distant cooperation should be feasible, and for versatility, exchange speed should be tended to. Because of these worries, not set in stone that the current structures should be improved to be used in casting a ballot framework.

SUMMARY

- **Cryptography:** Blockchain exchanges are validated and precise in light of calculations and cryptographic proof between the gatherings in question;
- **Immutability:** Any blockchain archives can't be changed or erased;
- **Provenance:** It alludes to the way that each exchange can be followed in the blockchain record;
- **Decentralization:** The whole appropriated data set might be open by all individuals from the blockchain network. An agreement calculation permits control of the framework, as displayed in the center interaction;
- **Anonymity:** A blockchain network member has created a location rather than a client distinguishing proof. It keeps up with obscurity, particularly in a blockchain public framework;
- **Transparency:** It implies being not able to control the blockchain network. It doesn't occur as it takes monstrous computational assets to delete the blockchain network.
- **Consensus:** An assortment of orders and associations to do blockchain processes.

VOTING BACKGROUND

Electronic democracy has been a space of exploration center for a long time by utilizing processing machines and hardware for projecting votes and creating top caliber and exact outcomes as per the feelings of the taking an interested citizen. Different endeavors have been taken on practically speaking to help the political race process.

At first PC counting framework permitted the citizen to choose inventions. Later on, those cards went through the most common way of filtering and counting at each surveying cell on a focal server. Direct Recording Electronic (DRE) casting ballot frameworks were set up later on which were respected and recognized enormously by the citizens regardless of the opposition from PC researchers. In case the democratic framework is surely known by the electors, the framework's ease of use can be expanded strikingly.

Protection

Keeping a singular's vote secret The framework uses cryptographic properties of blockchain to accomplish the security of an elector. All the more explicitly, as a citizen is enrolled into the framework, an elector hash is produced by blockchain which is the remarkable identifier of an elector into the blockchain and is shielded from abuse because of crash obstruction property of the cryptographic hash. Because of this, the discernibility of a vote is additionally non-unimportant along these lines ensuring the citizen when under coercion.

Privacy

With regards to the internet casting a ballot implies that nobody aside from the citizen realizes how a member has cast a ballot. Accomplishing this property predominantly depends on (at least one) of the accompanying methods: blind marks, homomorphic encryption, and blend networks [39]. A blind mark is a technique for marking information when the underwriter doesn't have a clue what they are marking. It is accomplished by utilizing a blinding capacity so that blinding and marking capacities are commutative–Blind(Sign(message)) = Sign(Blind(message)).

The requester blinds (applies blinding capacity to) their message and sends it for marking. After getting a mark for a dazed message, they utilize their insight into blinding boundaries to infer a mark for an unblinded message. Blind marks numerically keep anybody except the requester from connecting a dazed message and a comparing mark pair with an unblinded one.

Electronic Voting on Blockchain

This segment gives some foundation data on electronic democratic techniques. Electronic democracy is a democratic strategy wherein votes are recorded or counted utilizing the electronic gear. Electronic democracy is normally characterized as casting a ballot that is upheld by some electronic equipment and programming.

Such informalities ought to be skilled in supporting/executing different capacities, going from political decision arrangement through vote stockpiling. Stands at political race workplaces, workstations, and, all the more as of late, cell phones are largely instances of framework types. Citizen enrollment, confirmation, casting a ballot, and counting should be consolidated in the electronic democratic frameworks.

User Identity

As a username, blockchain uses aliases. This system doesn't give total protection and mystery. Since the exchanges are public, the client's character might be found by looking at and breaking down them. The blockchain's usefulness isn't appropriate to public decisions.

BRIEF DESCRIPTION

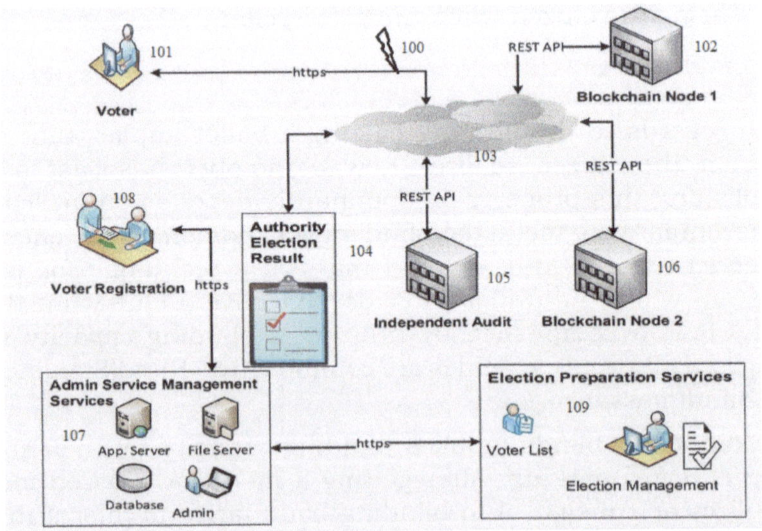

Fig. 1: Blockchain- based Voting Tracking and Management System, Flow Chart

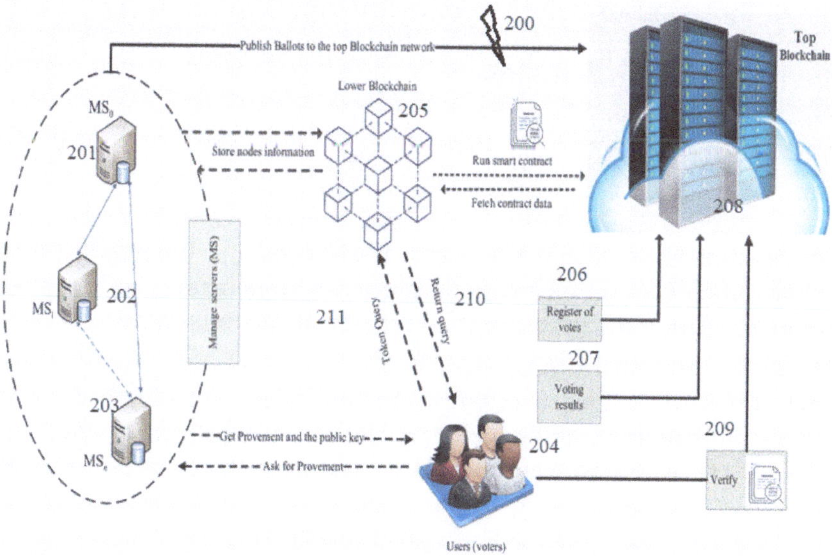

Fig. 2: Blockchain- based Voting Tracking and Management System, Block Diagram.

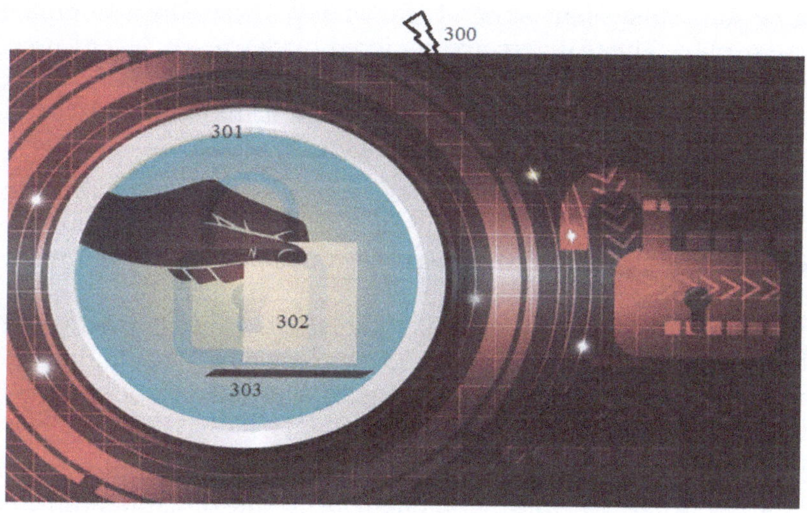

Fig. 3: Blockchain- based Voting Tracking and Management System

DESCRIPTION

The proposed e-casting ballot framework depends on the grounded Prêt à Voter e-casting a ballot approach distinguished in. The framework has been intended to help a democratic application in reality climate considering explicit prerequisites like security, qualification, comfort, receipt freeness, and certainty.

The proposed framework expects to accomplish secure advanced democracy without undermining its ease of use. Inside this unique circumstance, the framework is planned to utilize an online interface to work with client commitment with measures, for example, fingerprinting to secure against twofold democratic. With a reasonable need to manage the electors, electorates, and possibility for supporters, an easy-to-use manager interface is carried out to empower simple entry.

Moreover, the situation permits all electors equivalent privileges of cooperation and fosters a reasonable and solid contest among every one of the applicants while keeping the namelessness of the citizens saved. The cryptographic hash of the exchange (ID) is messaged to the citizen as proof that the vote has been cast which may, later on, be followed outside the premises of the supporters.

Detailed Description of the Layered Approach

The proposed e-casting ballot framework engineering is introduced in Fig. 1 and has been isolated into a few layers to accomplish the particular plan. These layers are depicted underneath; User Interaction and Front-end Security layer is liable for associating with a citizen (to help vote projecting capacities) and the chairman (to help capacities relating to managing the political decision process).

It embodies two key capacities for example verification and approval of the clients (electors and executives) to guarantee that the admittance to the framework is confined to real clients as per the predefined access control strategies. Various strategies can be applied to accomplish this capacity going from essential username/secret word to further developed like fingerprinting or iris acknowledgment. Subsequently, these are delivered explicitly to individual execution of the proposed engineering.

By and large, this layer fills in as the main resource with the clients and is answerable for approving client accreditations as administered by the framework explicit strategies.

Voting Process

We presently portray a common communication of a client with the proposed plot dependent on our present execution of the framework. Ordinarily, a citizen signs into the framework by giving his/her thumb impression. In case the match is found, the elector is then given a rundown of accessible applicants with the choice to choose them.

Despite what is generally expected, assuming the match is fruitless, any further access would be denied. This capacity is accomplished utilizing fitting execution of the validation component (fingerprinting for this situation) and predefined job-based admittance control the executives. Besides, it is likewise imagined that an elector is relegated to their particular voting demographic and this data is utilized to foster the rundown of competitors that a citizen can decide in favor of. The task of elector to a supporter is to deliver a disconnected interaction and along these lines out of the extent of this exploration.

Execution The execution of the proposed framework has been completed inside a controlled climate with an online application made to fill in as the front-end application empowering the clients to collaborate advantageously. This application is carried out through Java EE inside the NetBeans stage with a local Glassfish server utilized for facilitating the application.

Glassfish oversaw the server-side compartment for holding the application's EJBs and the information source. The application utilizes MySQL as the backend data set for the application and contains the information entered physically by an administrator, for example, the elector subtleties, body electorate subtleties, and the data about various ideological groups running for the political race. An application screen capture exhibiting the administrator capacity to see a rundown of qualified citizens is introduced.

Notwithstanding manual passages, the application likewise upholds bringing in information utilizing MS Excel spreadsheets to perform mass import considering the size of the information in certifiable democratic situations. We have utilized Multichip as the blockchain stage to make a private blockchain for this application which is utilized for recording the democratic exchanges. This decision is impacted by the convenience given by this stage and consequently, it was handily incorporated into our proposed engineering.

Electronic democratic or e-casting a ballot has been utilized in changing structures since the 1970s with principal benefits over invention-based frameworks like expanded proficiency and diminished mistakes. In any case, there remain difficulties to accomplish widespread reception of such frameworks particularly for working on their strength against likely blames.

Blockchain is a troublesome innovation of the current period and vows to work on the general versatility of e-casting a ballot framework.

The invention presents a top to bottom assessment of the plan which effectively shows its adequacy to accomplish a start to finish irrefutable e-casting ballot plot.

Regardless of these advantages, web-based democratic arrangements are seen with a lot of alerts since they present new dangers. A solitary weakness can prompt enormous scope controls of votes. Electronic democratic frameworks should be genuine, exact, safe, and advantageous when utilized for decisions. In any case, the reception might be restricted by potential issues related to electronic democratic frameworks.

The principal objective of this examination was to inspect the ebb and flow status of blockchain-based democratic exploration and web-based democratic frameworks and any connected troubles to foresee future turns of events. This review gives a theoretical depiction of the expected blockchain-based electronic democratic application and a prologue to the basic construction and qualities of the blockchain in association with electronic democracy.

- **Fig.1:** Explanation:100 is a Blockchain- based Voting Tracking and Management System Flow Chart, 101 is a input voter, 102and 106 is a blockchain node1,2, 103 is a cloud location, 104 is a result display unit, 105 is independent audit unit,107 is a admin Service management like server, file server, databased, admin, 108 is a voter registration, 109 is a election preparation services.
- **Fig. 2:** Explanation 200 is a Blockchain- based Voting Tracking and Management System Block Diagram, 201,202,203 is a MS server to manage the data baded,204 is a voter's user, 205 is a blockchain connection, 206 is a register of votes, 207 is a voting results, 208 is a final blockchain data display on global position, 209 is a verification process, 210,211 is a Query method and process two way.
- **Fig. 3:** Explanation 300 is a Blockchain- based Voting Tracking and Management System, 301,302,303 is a final image to tracking the voting system is a under controlled by authority under strangest digital security and result display immediately.

CLAIM

Our Invention "Blockchain-based Voting" This invention work to use advantages of blockchain, for example, cryptographic establishments and straightforwardness to accomplish a powerful plan for e-voting. The invented plot adjusts to the necessities for e-casting a ballot conspires and accomplishes start to finish undeniable nature. The invention presents subtleties of the proposed e-casting a ballot conspire alongside its execution utilizing the Multichain stage. Web-based democracy is a pattern that is acquiring energy in present-day culture. It can diminish hierarchical expenses and increment citizen turnout. It dispenses with the need to print voting from inventions or open surveying stations—citizens can cast a ballot from any place there is an Internet association. Blockchain innovation came into the ground to beat these issues and offers decentralized hubs for electronic democratic and is utilized to create electronic democratic frameworks mostly due to their start to finish confirmation benefits. This innovation is a delightful substitution for customary electronic democratic arrangements with disseminated, non-renouncement, and security insurance attributes. The accompanying article gives an outline of electronic democratic frameworks dependent on blockchain innovation. As a result of this review, it was found that blockchain frameworks might assist with settling a portion of the issues that currently plague political decision frameworks. Then again, the frequently referenced issues in blockchain applications are security assurance and exchange speed. For a maintainable blockchain-based electronic democratic framework, the security of distant cooperation should be feasible, and for versatility, exchange speed should be tended to. Because of these worries, not set in stone that the current structures should be improved to be used in casting a ballot framework.

- According to claim1, the Invention is to a "BMS- Vote Tracking: Blockchain-based Voting Tracking and Management System" This invention work to use advantages of blockchain, for example, cryptographic establishments and straightforwardness to accomplish a powerful plan for e-voting.
- According to claim1,2, the Invention is to possibly diminish hierarchical expenses and increment citizen turnout. It dispenses with the need to print voting from inventions or open surveying stations—citizens can cast a ballot from any place there is an Internet association.
- According to claim1,2,3, the Invention is a Blockchain innovation that came into the ground to beat these issues and offers decentralized hubs for electronic democracy, and is utilized to create electronic democratic frameworks mostly due to their start to finish confirmation benefits.

- According to claim1,2,3,4, the Invention is a result of this review, it was found that blockchain frameworks might assist with settling a portion of the issues that currently plague political decision frameworks. Then again, the frequently referenced issues in blockchain applications are security assurance and exchange speed.
- According to claim1,2,3,4, the Invention is to a maintainable blockchain-based electronic democratic framework, the security of distant cooperation should be feasible, and for versatility, exchange speed should be tended to. Because of these worries, not set in stone that the current structures should be improved to be used in casting a ballot framework.

REFERENCES

[1] Rahil Rezwan ; Huzaifa Ahmed ; M. R. N. Biplob ; S. M. Shuvo ; Md. Abdur Rahman., "Biometrically secured electronic voting machine" , IEEE Region 10 Humanitarian Technology Conference, 2021

[2] Himanshu Agarwal ; G. N. Pandey. "Online voting system for India based on AADHAAR ID" Paper available at: https://ieeexplore.ieee.org/document/6756265

[3] Vishal ; Vibhu Chinmay ; RisabhGarg; PoonamYadav "Online voting system linked with AADHAR" 3rd International Conference on Computing for Sustainable Global Development (INDIACom) Paper available at : https://ieeexplore.ieee.org/document/7724864

[4] Jameer Basha A Palanisamy V Purusothaman T ," Efficient multimodal biometric authentication using fast fingerprint verification and enhanced iris features" , Journal of Computer Science, 2020, vol 7, issues 5, pp 698 – 706.

[5] Gautam Srivastava, Ashutosh Dhar Dwivedi and Rajani Singh A Decentralized Voting Scheme using Blockchain Technology A Decentralized Voting Scheme using Blockchain Technology, Paper available at : www.insticc.org/Primoris/Resources/PaperPdf.ashx?idPaper=68819

[6] Friðrik Þ. Hjálmarsson, Gunnlaugur K. Hreiðarsson." Blockchain-Based E-Voting System", Paper available at: https://skemman.is/bitstream/1946/31161/1/Research-Paper-BBEVS.pdf

[7] Sagarshah , Qaishkanchwala , huaiqian mi . "Blockchain voting system", Paper available at: https://www.economist.com/sites/default/files/northeastern.pdf

[8] Kashif Mehboob Khan, Junaid Arshad, Muhammad Mubashir Khan, "Secure Digital Voting System based on Blockchain Technology" Paper Available at: https://pdfs.semanticscholar.org/c1f0/b096f9ce1b17bea2d39ee760aaede9829d29.pdf

[9] Gaby G. Dagher, Praneeth Babu, Marella, Matea Milojkovic, and Jordan Mohler. BroncoVote: Secure Voting System using Ethereum's Blockchain, Paper Available at: https://www.scitepress.org/papers/2018/66097/66097.pdf

[10] Nakamoto, S. (2008). Bitcoin: A peer-to-peer electronic cash system

[11] Ethereum Blog. (2018). On Public and Private Blockchains - Ethereum Blog. Available at: https://blog.ethereum.org/2015/08/07/ on-public-and-private-blockchains

CITATIONS

- Jain, J., Jain, A., Srivastava, S. K., Verma, C., Raboaca, M. S., & Illés, Z. (2022). Improved Security of E-Healthcare Images Using Hybridized Robust Zero-Watermarking and Hyper-Chaotic System along with RSA. Mathematics, 10(7), 1071.
- Jain, J., & Jain, A. (2022). Securing E-Healthcare Images Using an Efficient Image Encryption Model. Scientific Programming, 2022.
- Kumar, S., Shailu, A., Jain, A., & Moparthi, N. R. (2022). Enhanced Method of Object Tracing Using Extended Kalman Filter via Binary Search Algorithm. Journal of Information Technology Management, 14(Security and Resource Management challenges for Internet of Things), 180-199.
- Jain, A., Dwivedi, R. K., Alshazly, H., Kumar, A., Bourouis, S., & Kaur, M. (2022). Design and Simulation of Ring Network-on-Chip for Different Configured Nodes. CMC-COMPUTERS MATERIALS & CONTINUA, 71(2), 4085-4100.
- Sharma, S.K., Jain, A., Gupta, K., Prasad, D. and Singh, V., 2019. An internal schematic view and simulation of major diagonal mesh network-on-chip. Journal of Computational and Theoretical Nanoscience, 16(10), pp.4412-4417.

CHAPTER - 11

Electric Vehicle Charging Notification using IoT

Abhishek Yadav

abhi13.rao@gmail.com

Dr. Archana Shirbhate

Associate Professor, archana.shirbhate@rediffmail.com

B Koti Reddy

Scientific Officer, DAE, kotireddy1965@gmail.com

ABSTRACT

Our Invention "Advanced Method and Process Quick Electric Vehicle Charging" is an Electric Vehicles (EVs) are quickly turning into the heralds of vehicle innovation. First electric vehicles were disregarded in view of not having sufficient battery limit and in light of low effectiveness of their electric engines. Creating semiconductor and battery advances expanded the interest in the EVs. All things considered, current batteries actually have inadequate limit. Thus, vehicles should be re-energized at brief distances (around 150 km). Because of planned flight and appearance times EVs give off an impression of being more reasonable for city transports rather than customary autos. Because of right charging innovation and the accessibility of sustainable power for electric transports, the urban communities have less commotion and CO2 emanations. The energy utilization of gas powered motors is higher than of the electric engines. In this paper, studies on the business electric vehicle charging strategies will be checked on and the module charging cycles will be depicted exhaustively.

This study endeavors to address the inquiries of how module charging process correspondence has performed between the EV and Electric Vehicle Supply Equipment (EVSE). Toward the start of the charging system, EVSE checks the disengagement of the great voltage framework and it continues announcing this worth during the entire charging time. The electric power is gotten from the network by the rectifier and it is conveyed to the DC/DC converter. The DC high voltage can be created by 3-stage AC. The DC power is associated with a channel in the wake of being changed over by 3-stage rectifiers; the result of the channel is associated with a DC/DC converter lastly went through one more channel to the vehicle battery.

Key: *Electric, Vehicle, Charging, Notification, IoT, Electric, Vehicle, Supply, Equipment.*

FIELD

Our Invention is related to a Advanced Method and Process Quick Electric Vehicle Charging.

BACKGROUND

For a dependable future for EVs in the business area, a few significant issues ought to be settled. The driving scope of battery-electric transports is a limitation and the charging system requires a specific measure of time. Further developing battery innovation gave a higher reach to electric vehicles, however the span of the charging system expanded.

To dispose of long charging times, fast charging innovation has been created. Nonetheless, speedy charging innovation has a few adverse consequences on the lattice. Transient pinnacle upsides of momentary charging power might request inordinate power from the framework.

Many investigations are broke down with regards to changes happening on the lattice, which brought about separating the charging choices of electric vehicles into two subsections. This paper concentrates on the Mode 4 charging framework and the Combined Charging System (CCS).

OBJECTIVES

- The objective of the invention is to provide a "Advanced Method and Process Quick Electric Vehicle Charging" is an Electric Vehicles (EVs) are quickly turning into the heralds of vehicle innovation. First electric vehicles were disregarded in view of not having sufficient battery limit and in light of low effectiveness of their electric engines.

- The other objective of the invention is to provide a Creating semiconductor and battery advances expanded the interest in the EVs. All things considered, current batteries actually have inadequate limit. Thus, vehicles should be re-energized at brief distances (around 150 km).
- The other objective of the invention is to provide a Because of planned flight and appearance times EVs give off an impression of being more reasonable for city transports rather than customary autos. Because of right charging innovation and the accessibility of sustainable power for electric transports, the urban communities have less commotion and CO2 emanations.
- The other objective of the invention is to provide a energy utilization of gas powered motors is higher than of the electric engines. In this paper, studies on the business electric vehicle charging strategies will be checked on and the module charging cycles will be depicted exhaustively.
- The other objective of the invention is to provide a endeavors to address the inquiries of how module charging process correspondence has performed between the EV and Electric Vehicle Supply Equipment (EVSE).
- The other objective of the invention is to provide a toward the start of the charging system, EVSE checks the disengagement of the great voltage framework and it continues announcing this worth during the entire charging time.
- The other objective of the invention is to provide a electric power is gotten from the network by the rectifier and it is conveyed to the DC/DC converter. The DC high voltage can be created by 3-stage AC.
- The other objective of the invention is to provide a DC power is associated with a channel in the wake of being changed over by 3-stage rectifiers; the result of the channel is associated with a DC/DC converter lastly went through one more channel to the vehicle battery.

SUMMARY

Contingent upon the charging power, EVs can be associated with the matrix straightforwardly or by means of a piece of hardware.

The hardware is otherwise called EVSE. EVSE is an expansive term characterizing all the establishing and non-establishing gear, electric vehicle connectors, connection fittings and any remaining sorts of frill required for charging the EVs. The main component is the two-way correspondence between the charging unit and the vehicle.

There are two worldwide connector types for Mode-4 charging. Chademo is the primary framework for DC module charging process and Combined Charging System (CCS) is another variety that followed it. Control and correspondence conventions vary from one another.

The CCS utilizes PWM and Power Line Communication (PLC) over the control pin and for precise situating of the charger plug it likewise utilizes the closeness pin, while CHAdeMO utilizes the Controller Area Network (CAN) Bus correspondence for control and correspondence through seven pins.

BRIEF DESCRIPTION OF THE DIAGRAM

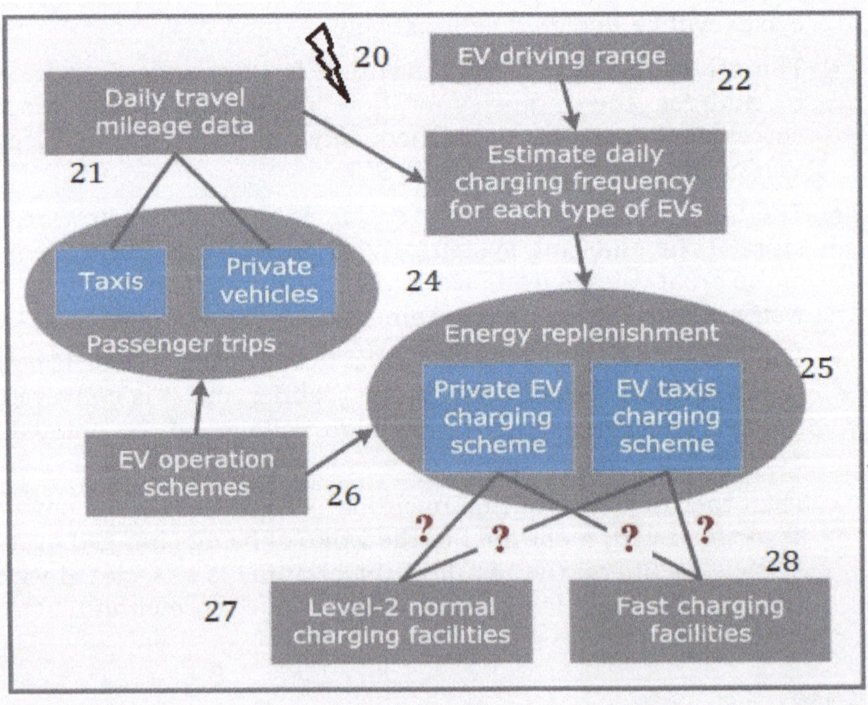

Fig. 1: Advanced Method and Process Quick Electric Vehicle Charging, Process

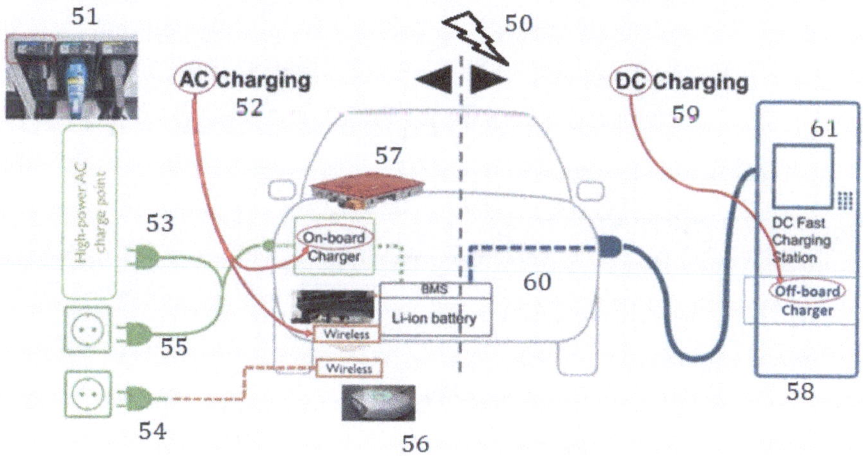

Fig. 2: Advanced Method and Process Quick Electric Vehicle Charging, Connective

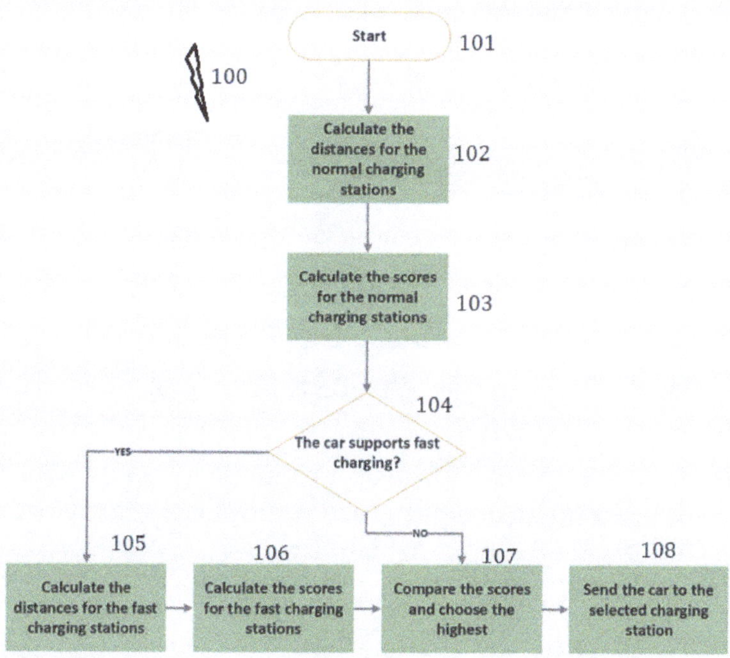

Fig. 3: Advanced Method and Process Quick Electric Vehicle Charging, Flow Chart.

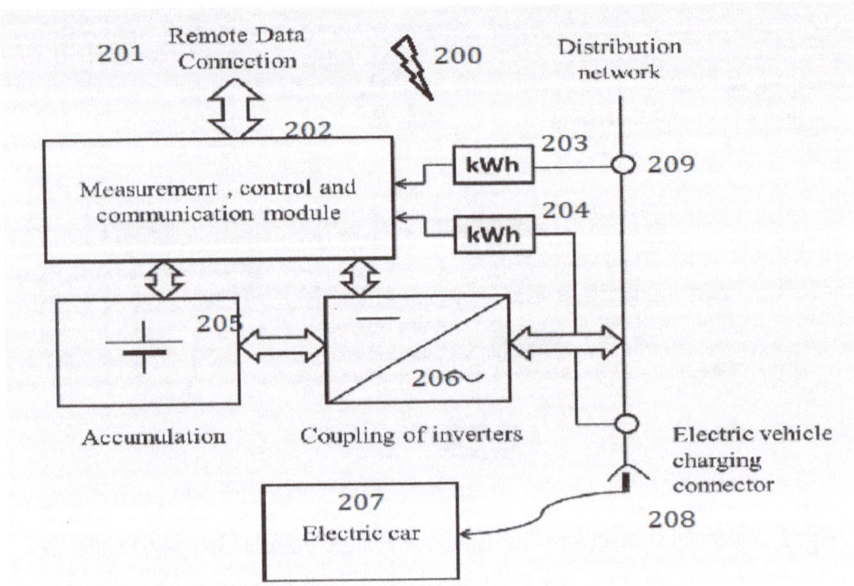

Fig. 4: Advanced Method and Process Quick Electric Vehicle Charging, Block Diagram.

Fig. 5: Advanced Method and Process Quick Electric Vehicle Charging

DESCRIPTION

CCS (Combined Charging System) The Society of Automotive Engineers (SAE), the European Automobile Manufacturers Association (ACEA) and a few car producers (Audi, BMW, Daimler, Ford, General Motors, Porsche, and Volkswagen) have fostered the CCS as another worldwide EV charging framework.

The CCS has two connector types, CCS Type 1 and CCS Type 2. Because of these connectors, EVs can be energized to 50-350 kilowatts. CCS Type 1 bay was created by Kiyoshi Osawa. CCS Type 1 and CCS Type 2 bays are shown in Figures 4a and 4b.

The principle contrast between these bay sorts is the stage kind of AC. CCS Type 1 framework can be utilized in uses of EV charging through mono-stage and CCS Type 2 for three stages. CCS Type 1 framework is likewise legitimate for US electric vehicles, while CCS Type 2 is just appropriate for Europe.

Quick charging through pantograph DC quick charging framework by means of pantograph can be considered as an improved state for the electric trolleybuses, for which the energy dissemination framework disturbs metropolitan preparation.

The framework is reasonable for city transports. Trolleybuses have been stimulated from long electric conveyance lines yet on DC David Steen and Le Anh Tuan concentrated on the impacts of electric transports on DC quick charging stations in the metropolitan matrix and furthermore they explored the greatest power interest of quick charging stations.

Robert Prohaska et al. concentrated on electric transports on the transport line 291 between La Verne and Pomona. Twelve BE35 model electric transports were purchased from Proterra Inc. in 2013 for the 291 transport line. For testing, the vehicles were driven on a 142,605 km range course.

The normal temperatures are taken from the National Oceanic and Atmospheric Administration (NOAA) for the Ontario International Airport, which is around 15 km away from the electric transports charged. The proficiency of vehicles is greatest when the surrounding temperature is near room temperature.

The month to month proficiency of the transports vacillates comparative with normal temperatures. Whenever the encompassing temperatures are higher, the transports require the utilization of cooling while they require radiators when the temperatures drop.

The electric power is changed over into attractive power by groundwork loops, then, at that point, the optional curl gathers the created electromagnetic field energy and it is changed into the electric power, then, at that point, EVs are charged. This occasion clarifies the primary idea of transformers. To keep away from electrical harms, the curls have a similar resounding recurrence. By sending the flow to batteries of the electric vehicle, the charging system is started.

The inductive charging process has effectively been utilized for little electronic apparatuses (cell phones, toothbrushes, and so forth) Due to the shortfall of electrical contacts, the transmitter and recipient of the IPT framework are free. This include altogether builds the security and decreases the chance of defacing.

Georgiy Babat fostered the primary utilization of the IPT framework on the EV in 1943. The effectiveness of the application was just 4%. The primary remote charging innovation was fixed; the framework is likewise called the static inductive charging framework. For charging EVs during travel, the powerful remote charging framework was created.

In an IPT framework, it is indispensable to trade data between the charger side and the EV side to give criticism. The correspondence can be partitioned into whether the sign is tweaked on the electrical cables or then again assuming it applies a different recurrence band. Leandros A. Maglaras et al. concentrated on the unique remote charging framework.

They declared Mobile Energy Disseminators (MEDs), another interaction that can make it workable for EVs to broaden their driving reach in the city. The venture utilizes between vehicle correspondences. Specific hubs, similar to business vehicles, can proceed as energy sources to EVs. Drugs utilize the IPT framework to charge the destitute EVs. An open point of interaction standard for low power IPT framework was created by the Wireless Power Consortium (WPC).

Andrea Gil et al. concentrated on the suitable specialized techniques for IPT frameworks. As per the outcomes, low-rate remote Personal Area Network (PAN) gear like ZigBee might be utilized for fixed IPT frameworks. Be that as it may, customary correspondence PANs are deficient for dynamic IPT on account of the reach.

The looks at remote advances utilized by Communication Access for Land Mobiles (CALM). Satellite has the best inclusion include. Be that as it may, it is unseemly for IPT charging due to the inertness issue. A similar inertness issue can be seen in FM radio and WiMAX.

POWER ELECTRONIC SYSTEMS

The power electronic circuits of the EV charging frameworks are performed by an AC/DC rectifier and a DC/DC converter. This geography is legitimate for a wide range of charging strategies (module charging, pantograph charging, and, surprisingly, remote charging).

As per IEC 15118-2:3 guidelines, the result voltage of EVSE should be controlled. During the pre-charging province of EVSE, the electric vehicle gets under 2 A flow and a specific voltage. In the event that the voltage distinction between the electric vehicle and EVSE is under 20 V, the charging system can be started.

Toward the start of the charging system, EVSE checks the seclusion of the great voltage framework and it continues detailing this worth during the entire charging time. The electric power is gotten from the lattice by the rectifier and it is conveyed to the DC/DC converter. The DC high voltage can be delivered by 3-stage AC.

The DC power is associated with a channel in the wake of being changed over by 3-stage rectifiers; the result of the channel is associated with a DC/DC converter lastly went through one more channel to the vehicle battery.

The paper presents the impacts of EVSE on the matrix as well as the other way around, the impacts of semiconductors and inverter types on the effectiveness of EVSE, the presentation of the correspondence channel. Notwithstanding these investigations, beneficial clarifications are made for getting the framework.

At the point when the benefits and detriments are recorded by the motivation behind various kinds of converters, it has been tracked down that controlled AC/DC rectifiers are suggested for the EVSE, and DC/DC bidirectional DAB converters are suggested for wide-range charging.

Because of the consonant impacts of rectifiers to the matrix, utilizing LCL channel and to keep away from the temperamental activity of the channel, it is suggested that an extra resistor should be carried out on the power circuit of EVSE. It is likewise reasonable to utilize the PLC signals from the PWM channel (control pilot) when the consequences of the activity for the correspondence channel are assessed.

As per IEC 15118, standard, PLC framework is chosen for the significant level correspondence convention. Along these lines, a V2G energy move framework can be utilized. It is additionally prescribed to utilize PID regulators to stay away from glitch or disappointment in the framework.

RESEARCH CLAIM

Our Invention "Advanced Method and Process Quick Electric Vehicle Charging" is an Electric Vehicles (EVs) are quickly turning into the heralds of vehicle innovation. First electric vehicles were disregarded in view of not having sufficient battery limit and in light of low effectiveness of their electric engines. Creating semiconductor and battery advances expanded the interest in the EVs. All things considered, current batteries actually have inadequate limit. Thus, vehicles should be re-energized at brief distances (around 150 km). Because of planned flight and appearance times EVs give off an impression of being more reasonable for city transports rather than customary autos. Because of right charging innovation and the accessibility of sustainable power for electric transports, the urban communities have less commotion and CO2 emanations. The energy utilization of gas powered motors is higher than of the electric engines. In this paper, studies on the business electric vehicle charging strategies will be checked on and the module charging cycles will be depicted exhaustively. This study endeavors to address the inquiries of how module charging process correspondence has performed between the EV and Electric Vehicle Supply Equipment (EVSE). Toward the start of the charging system, EVSE checks the disengagement of the great voltage framework and it continues announcing this worth during the entire charging time. The electric power is gotten from the network by the rectifier and it is conveyed to the DC/DC converter. The DC high voltage can be created by 3-stage AC. The DC power is associated with a channel in the wake of being changed over by 3-stage rectifiers; the result of the channel is associated with a DC/DC converter lastly went through one more channel to the vehicle battery.

- According to claim1# the invention is to a "Method and Process Quick Electric Vehicle Charging" is an Electric Vehicles (EVs) are quickly turning into the heralds of vehicle innovation. First electric vehicles were disregarded in view of not having sufficient battery limit and in light of low effectiveness of their electric engines.
- According to claim1,2# the invention is to a Creating semiconductor and battery advances expanded the interest in the EVs. All things considered, current batteries actually have

inadequate limit. Thus, vehicles should be re-energized at brief distances (around 150 km).

- According to claim1,2,3# the invention is to a Because of planned flight and appearance times EVs give off an impression of being more reasonable for city transports rather than customary autos. Because of right charging innovation and the accessibility of sustainable power for electric transports, the urban communities have less commotion and CO2 emanations.
- According to claim1,2,4# the invention is to a energy utilization of gas powered motors is higher than of the electric engines. In this paper, studies on the business electric vehicle charging strategies will be checked on and the module charging cycles will be depicted exhaustively.
- According to claim1,2,4,5# the invention is to a endeavors to address the inquiries of how module charging process correspondence has performed between the EV and Electric Vehicle Supply Equipment (EVSE).
- According to claim1, 2,3,5,6# the invention is to a toward the start of the charging system, EVSE checks the disengagement of the great voltage framework and it continues announcing this worth during the entire charging time.
- According to claim1,2,4,7# the invention is to an electric power is gotten from the network by the rectifier and it is conveyed to the DC/DC converter. The DC high voltage can be created by 3-stage AC.
- According to claim1, 2,3,5,8# the invention is to a DC power is associated with a channel in the wake of being changed over by 3-stage rectifiers; the result of the channel is associated with a DC/DC converter lastly went through one more channel to the vehicle battery.

REFERENCES

[1] S. Yonghua, Y. Yuexi, H. Zechun, "Present Status and Development Pattern of Batteries for Electric Vehicles," Force Framework Innovation, Vol. 35, No. 4, pp. 1-7, 2021.

[2] L. Xiaokang, Z. Qionghua, H. Kui, S. Yuehong, "Battery the board framework for electric vehicles," J.Huazhong Univ. Of Sci. and Tech. (Nature Science Release). Vol. 35, No. 8, pp. 83-86, 2020.

[3] C. Piao, Q. Liu, Z. Huang, C. Cho, and X. Shu, "VRLA Battery The executives Framework Dependent on LIN Transport for Electric Vehicle," Trend setting innovation in Instructing, AISC163, pp. 753-763,2022

[4] J. Chatzakis, K. Kalaitzakis, N. C. Voulgaris and S. N. Lunacies, "Planning another summed up battery the framework", IEEE Trans. Ind. Electron. Vol. 50, No. 5, pp. 990 - 999, 2019.

[5] D. S. Suresh, Sekar R, Mohamed ShafiullaS.,"Battery Checking framework Dependent on PLC",Worldwide Diary of Science and Exploration, vol. 3 issue 6. pp. 128-133, 2012.

[6] A. Sardar, H. Naseer, E. Qazi, and W. Ali "Shrewd Networks Wide Region Checking Framework forUPS Batteries over GSM" second Worldwide Multidisplinary Meeting For Better Pakistan Vol.1, pp. 159-158, May 2020, 2015.7. C. Hommalai and S. Khomfoi "Battery Checking Framework by Identifying Dead Battery Cells", Worldwide Diary of Science and Exploration, Vol.1, pp. 5-15, 2020.

[7] A. S. Dhotre, S. S. Gavasane, A. R. Patil, and T. Nadu, "Programmed Battery Charging Utilizing Battery Wellbeing Discovery" Worldwide Diary of Designing and Innovation. Inventive science vol. 1, no. 5, pp. 486–490, 2014.

[8] S. A. Mathew, R. Prakash, and P. C. John "A shrewd remote battery observing framework for electric vehicles," Int. Conf. Intel. Syst. Des. Appl. ISDA, pp. 189–193, 2012.

[9] S. Bacquet, M. Maman, "Radio recurrence correspondences for shrewd cells in battery pack for electric vehicle", Electric Vehicle Meeting (IEVC) 2014 IEEE Global, pp. 1-4, 2014.

CITATIONS

- Jain, A., Dwivedi, R. K., Alshazly, H., Kumar, A., Bourouis, S., & Kaur, M. (2022). Design and Simulation of Ring Network-on-Chip for Different Configured Nodes. CMC-COMPUTERS MATERIALS & CONTINUA, 71(2), 4085-4100.
- Jain, A., Kumar, A. and Sharma, S., 2015. Comparative Design and Analysis of Mesh, Torus and Ring NoC. Procedia Computer Science, 48, pp.330-337.
- Jain, A., Dwivedi, R., Kumar, A. and Sharma, S., 2017. Scalable design and synthesis of 3D mesh network on chip. In Proceeding of International Conference on Intelligent Communication, Control and Devices (pp. 661-666). Springer, Singapore.
- Jain, A., Gahlot, A.K., Dwivedi, R., Kumar, A. and Sharma, S.K., 2018. Fat Tree NoC Design and Synthesis. In Intelligent Communication, Control and Devices (pp. 1749-1756). Springer, Singapore.
- Jain, A., Dwivedi, R., Kumar, A. and Sharma, S., 2017. Scalable design and synthesis of 3D mesh network on chip. In Proceeding of International Conference on Intelligent Communication, Control and Devices (pp. 661-666). Springer, Singapore.

Chapter - 12

Monthly Salary Notification using IoT

Basudeo Singh Roohani, Ankur Rana, Prof. (Dr.) Vivek Kumar
Quantum University, Dehradun Highway, Mandawar, Roorkee, Uttarakhand 247167, India.

ABSTRACT

Our Invention "Compare the Dates in a Month with the Monthly Salary using Regression Method" is a break down the connection between private pay and the typical number of hours worked each week. Work market financial matters infer a positive connection between hours worked and higher pay, yet in addition recommend that beyond a particular pay edge, people will more often than not work less hours. The connection between hours worked and individual pay is demonstrated utilizing information from the United States Census Bureau's American Community Survey from 2013 and 2019. Straightforward relapse investigation tried the log of changed individual pay against hours worked, and the different relapse extended this examination to incorporate gas utility costs, number of laborers in family, food stamp help, and number of people in family as factors. The basic relapse model exhibits that individual pay has a positive relationship with hours worked, while the numerous relapse model shows that this impact reduces as pay level increments. A mechanized strategy and framework gauges pay of a singular advance candidate utilizing credit authority data and advance ascribes. The strategy and framework can utilize the credit department and advance data to adjust a candidate's obligation trouble in situations where such data is not promptly accessible or is mysterious. The strategy and framework can naturally check pay for candidates who decide to express their pay in lieu of giving sufficient documentation. Further, the strategy and framework can be material to any retail loaning business including, however not restricted to, contract, vehicle advance, and Mastercards, where credit authority data shapes a piece of the information assortment process and is accessible alongside candidate's data. The past area utilizes a graphical way to deal with assess the exhibition of our straight model, which can in some cases be very abstract in our judgment. Here we will utilize a few quantitative strategies to acquire a more exact presentation assessment of our straight model.

FIELD OF THE INVENTION

Our Invention is related to a Compare the Dates in a Month with the Monthly Salary using Regression Method.

BACKGROUND OF THE INVENTION

The connection among pay and hours worked is an issue of extraordinary useful and, for certain individuals, maybe even upright importance. Largely, than those of low pay. Maybe the generously compensated decide to work more in light of the fact that each extra hour worked is exceptionally rewarding, subsequently procuring even more in light of the fact that they work more. Then again, they may not function as much since they can make to the point of satisfying their requirements in a more limited measure of time.

Maybe those acquiring low pay rates should work more since that is the best way to procure a satisfactory pay. It is likewise conceivable that those acquiring less work less because they do not consider it advantageous to invest extra energy working for low wages. The solution to this inquiry can be reduced to whether or not higher wages boost individuals to work less or work more. The hypothetical standards portraying these motivations – the pay impact and the replacement impact – will be examined in more detail in the accompanying area.

Prior to delving into such subtleties, we present the accompanying theory. The pay impact proposes that higher pay makes individuals work less hours. The replacement impact proposes that higher pay makes individuals work more hours. Our speculation is that the pay impact is for the most part more grounded and, accordingly, higher pay should prompt less hours worked.

Does our compensation truly develop with our long periods of working experience? While this can be addressed dependent on our overall understandings of the gig market, we will utilize a straightforward information driven way to deal with check the reality. We will utilize straight relapse to show the connection between how much compensation with the long periods of working experience.

Direct relapse is a model that expects a straight connection between a logical variable (X) and a reaction variable (y). We can anticipate the worth of y dependent on the worth of X. In our setting here, the assessed compensation will be our reaction variable (y) since it is our objective anticipated worth and the long periods of working experience will be our logical variable (X).

Whenever we have characterized the illustrative variable and reaction variable for our situation, we will utilize Python to fabricate a straight relapse model to address our inquiry.

PRIOR ART SEARCH

Expressed pay credits might be seen to be less secure than full documentation advances. Without a sufficient check process, the bank chances that a few candidates might exaggerate their pay to accomplish lower relationship of outstanding debt to take home pay, a vital determinant of installment capacity in the guaranteeing system, to get endorsement for a specific advance. Accordingly, candidates expressing their pay might think twice about higher rates, bigger up front installments, higher financial assessment necessities, or a blend thereof.

According to the moneylender's viewpoint, such tradeoffs may not legitimize the harmony among hazard and compensation for expressed pay credits. According to the candidates' viewpoint, higher rates and bigger initial investments are not attractive for the people who really expressed their real pay and decided on the expressed pay program to improve on the credit handling technique or to keep up with their security.

Regular pay assessment frameworks are utilized in the fields of financial aspects and sociology, just as by the U.S. government. In any case, these frameworks commonly don't gauge a singular's pay and don't use past credit and hazard execution got from credit authority ascribes or a candidate's advance data. Different organizations of the U.S. government have created various philosophies for assessing middle pay with the end goal of an area pay statistics, lodging reasonableness, or territorial neediness levels.

In one ordinary framework, the middle family pay for a little district was assessed as an element of different factors taken from regulatory records. Albeit this strategy straightforwardly connects with pay assessment, it does not mean pay assessment for a person.

In another non-closely resembling customary framework, a pay assessment strategy associates instruction levels with family pay, which is not appropriate in retail advance handling. In this way, it is alluring to have a strategy and a framework that gauges a candidate's pay for a retail loaning program by utilizing credit agency and advance ascribes.

OBJECTIVES OF THE INVENTION

- The objective of the invention is to provide a "Compare the Dates in a Month with the Monthly Salary using Regression Method" is a breaks down the connection between private pay and the typical number of hours worked each week. Work market financial matters infer a positive connection between hours worked and higher pay, yet in addition recommend that beyond a particular pay edge, people will more often than not work less hours.
- The other objective of the invention is to provide a connection between hours worked and individual pay is demonstrated utilizing information from the United States Census Bureau's American Community Survey from 2013 and 2019. Straightforward relapse investigation tried the log of changed individual pay against hours worked, and the different relapse extended this examination to incorporate gas utility costs, number of laborers in family, food stamp help, and number of people in family as factors.
- The other objective of the invention is to provide a basic relapse model exhibits that individual pay has a positive relationship with hours worked, while the numerous relapse model shows that this impact reduces as pay level increments. A mechanized strategy and framework gauges pay of a singular advance candidate utilizing credit authority data and advance ascribes.
- The other objective of the invention is to provide a strategy and framework can utilize the credit department and advance data to adjust a candidate's obligation trouble in situations where such data is not promptly accessible or is mysterious. The strategy and framework can naturally check pay for candidates who decide to express their pay in lieu of giving sufficient documentation.
- The other objective of the invention is to provide a strategy and framework can be material to any retail loaning business

including, however not restricted to, contract, vehicle advance, and Mastercards, where credit authority data shapes a piece of the information assortment process and is accessible alongside candidate's data.

- The other objective of the invention is to provide a past area utilizes a graphical way to deal with assess the exhibition of our straight model, which can in some cases be very abstract in our judgment. Here we will utilize a few quantitative strategies to acquire a more exact presentation assessment of our straight model.

SUMMARY OF THE INVENTION

A mechanized strategy and framework for assessing pay of a singular advance candidate utilizes credit agency data and advance ascribes. The strategy and framework can utilize the credit authority and advance data to align a candidate's obligation trouble in situations where such data isn't promptly accessible or is mysterious.

The strategy and framework can naturally confirm pay for candidates who decide to express their pay in lieu of giving satisfactory documentation. Further, the technique and framework can be relevant to any retail loaning business including, yet not restricted to, contract, vehicle advance, and charge cards, where credit agency data shapes a piece of the information assortment process and is accessible alongside candidate's data.

It is positive that the strategy and framework extricate the significant data from credit department and advance data to assess a candidate's actual pay. Further, it is attractive to furnish banks with a choice to broaden a candidate the advantage of beneficial evaluating in an expressed pay advance program dependent on a correlation between the candidate's expressed pay and the assessed pay.

The strategy and framework depicted in these utilization procedures to choose most prescient factors from an enormous pool of up-and-comers, tidy up the possible exceptions/blunders among an informational collection, and concentrates the important data from the competitor indicators to construct a last model to appraise the candidate's pay and also the boundaries of a multivariate versatile relapse splines ("MARS") based expectation framework are assessed from a data set comprising of borrower data on full-documentation credit customers, where the genuine pay are known and have been checked.

Improvement/wait/out-of-time approvals alongside bootstrap resampling procedures give a model that endeavors to limit the mistake between genuine pay and anticipated pay. Besides, a careful and efficient examination is performed between expressed obligation proportion, i.e., obligation trouble determined from the candidate's expressed pay, and anticipated obligation proportion, i.e., obligation trouble determined from the assessed pay.

Loading Data

Firstly, we will use the *Python Pandas* library to read our CSV data.

```
import pandas as pd
import numpy as np
from sklearn.model_selection import train_
test_split
from sklearn import linear_model
import sklearn.metrics as sm
import matplotlib.pyplot as plt

data = pd.read_csv("Data/Salary_Data.csv")
X = data['YearsExperience']
y = data['Salary']
```

Dividing information into a preparation set and a test set. Before building a straight model, we really want to set up a preparation set and a test set (part of standard system in an AI work process). The preparation set will be utilized to prepare the model though the test set will be utilized to survey the presentation of the prepared model in anticipating the outcome from inconspicuous information.

We can utilize the Python scikit-learn train_test_split capacity to haphazardly divide our information into a preparation and test set

Set separated 30% of the whole dataset as the test set and allocate the preparation set and test set into four factors, individually.

The information comes from the United States Census Bureau's American Community Survey. We have consolidated the individual and family informational indexes from 2013 to make one huge informational collection. This was conceivable because of chronic numbers coordinating people with their family.

A couple of factors have been added or altered, as point-by-point beneath in the variable depictions. We clarify every factor and its significance beneath. A table containing elucidating measurements for every factor joins our variable clarifications. The reliant variable is wkhp, the autonomous variable from the straightforward relapse is ladjpincp, and any remaining factors will be utilized in the numerous Regression.

Predictive Analytics

- The initial step is to get preparing informational collection and test informational collection. We should utilize 80% as preparing information and staying 20% for testing.
- Blue dabs addresses the test information showing connection between house cost and square feet of residing region. Green line shows the forecast of the home value (subordinate variable) for given square feet utilizing the "sqft_model" direct relapse model we constructed.

BRIEF DESCRIPTION OF THE DIAGRAM

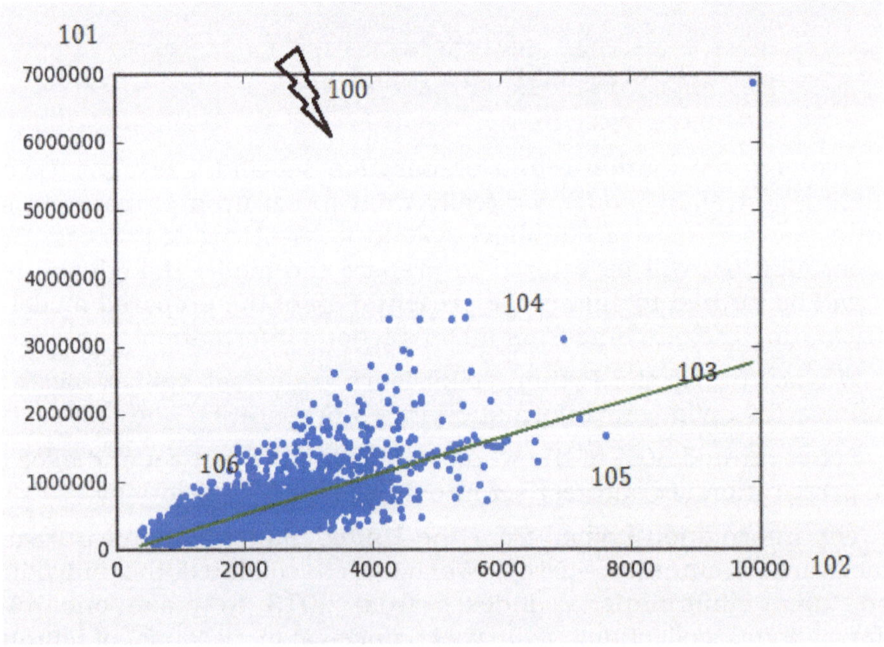

Fig. 1: Predictive Analytics

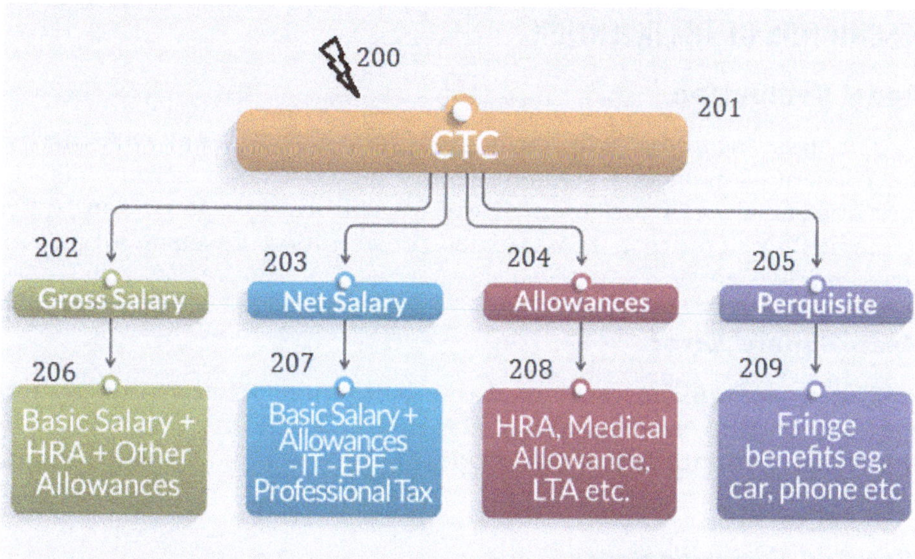

Fig. 2: Compare the Dates in a Month with the Monthly Salary, Flow Chart.

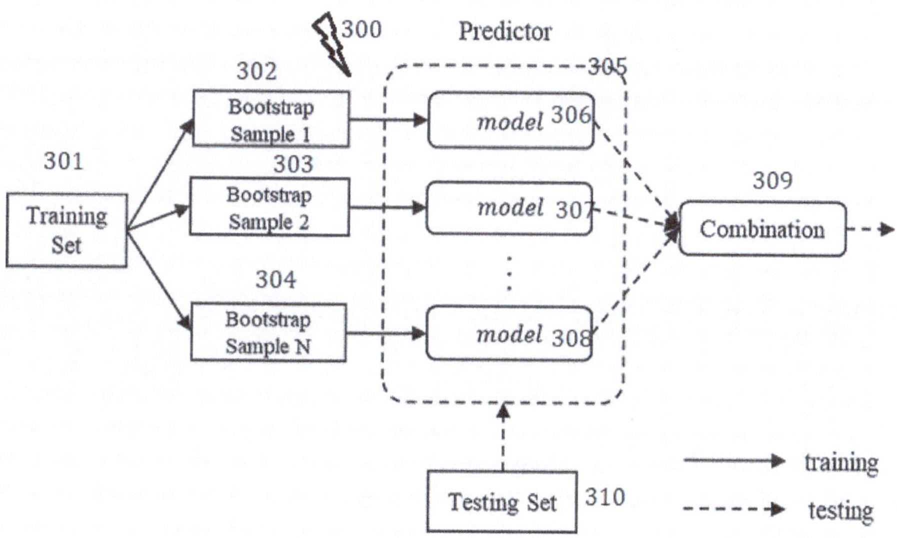

Fig. 3: Salary Predictor using regression method

DESCRIPTION OF THE INVENTION

Model Evaluation

The past segment utilizes a graphical way to deal with assess the presentation of our straight model which can in some cases be very abstract in our judgment. Here we will utilize a few quantitative strategies to acquire a more exact presentation assessment of our direct model.

Mean Square Error

The normal of the squares of the distinction between the genuine qualities and the anticipated qualities. The lower the distinction the better the presentation of the model. This is a typical measurement utilized for relapse investigation.

Clarified Variance Score

An estimation to inspect how well a model can deal with the variety of qualities in the dataset. A score of 1.0 is the ideal score.

R2 Score

An estimation to analyze how well our model can foresee values dependent on the test set (obscure examples). The ideal score is 1.0

Mid Variable: This is a double factor addressing laborers with an individual pay more noteworthy than $20,599 however not exactly or equivalent to $104,096. Since financial hypothesis on the contending replacement and pay impacts recommends that the impact of pay on hours worked may change as pay expands, we have included faker factors to represent low, center, and top level salary. This variable, on the off chance that it has a worth of 1, signifies a person of center individual pay. If zero, the individual is either in the top or base pay quintile.

High Variable: The double factor high demonstrates whether individual pay in the beyond a year is more prominent than $104,096. As expressed above, we try to represent how the impact of pay on hours worked changes with pay level. This variable, assuming it has a worth of 1, means a person of high private pay. The decision of these particular pay esteems (for both the high and mid factors) depends on the dispersion of pay in the United States. The lower fifth quintile lies

9 beneath \$20,600, the second through fourth quintiles range from \$20,600 to \$104,096, and the fifth quintile comprises of all salaries bigger than \$104,096 (5)

Npf Variable: The quantity of people in the specialist's family is addressed by the variable npf. The quantity of individuals in an individual's family might influence how much time the person decides to work. Maybe individuals with huge families should work more to accommodate a huge gathering of individuals. Then again, somebody with an enormous family could put an uncommonly high worth on recreation and time with family, which could lead them to work less. This variable just records for family families, and it gives information in the reach 2 – 20.

Fs Variable: The variable fs is a twofold worth showing whether or not an individual gets food stamps. An individual getting food stamps might work longer to try not to need government support. On the other hand, an individual getting food stamps might work less because the public authority gives a portion of their necessities. The informational index initially gave up sides of 1 to those getting help and 2 for those not getting food stamps. To be predictable with the generally involved upsides of 0 and 1 for parallel factors, fs was recoded so those not getting food stamps have a worth of 0.

wif Variable: The quantity of laborers in the family during the most recent a year is addressed by the variable wif. In families with an enormous number of laborers, not every individual specialist might need to fill in however many hours as somebody who is the main laborer in their family. This variable just records for family families and gives information in the reach 0 – 3 (with families with multiple laborers being covered at 3).

lgasp Variable: The log of the month-to-month cost of gas utilities is addressed by lgasp. Individuals might decide to work more in the event that the expense of utilities is high. We got this variable by taking the log of the variable wheeze (month-to-month cost of gas utilities) gave in the first informational index.

The first informational index gives esteems to the scope of \$4 – \$999 each month. It relegated upsides of 1¬3 for subjective results for conditions like remembering gas utilities for lease, included them power, or not utilizing gas utilities. We have allotted those qualities to 1 (it ought to be noticed that log(1) = 0) on the grounds that the individual isn't taking care of an autonomous gas bill.

We picked the log of the variable because a rate change will be more significant than a flat out change (for example 10% expansion is significant, while a $10 increment may not be enormous).

WE CLAIMS

Our Invention "Compare the Dates in a Month with the Monthly Salary using Regression Method" is a break down the connection between private pay and the typical number of hours worked each week. Work market financial matters infer a positive connection between hours worked and higher pay, yet in addition recommend that beyond a particular pay edge, people will more often than not work less hours. The connection between hours worked and individual pay is demonstrated utilizing information from the United States Census Bureau's American Community Survey from 2013 and 2019. Straightforward relapse investigation tried the log of changed individual pay against hours worked, and the different relapse extended this examination to incorporate gas utility costs, number of laborers in family, food stamp help, and number of people in family as factors. The basic relapse model exhibits that individual pay has a positive relationship with hours worked, while the numerous relapse model shows that this impact reduces as pay level increments. A mechanized strategy and framework gauges pay of a singular advance candidate utilizing credit authority data and advance ascribes. The strategy and framework can utilize the credit department and advance data to adjust a candidate's obligation trouble in situations where such data is not promptly accessible or is mysterious. The strategy and framework can naturally check pay for candidates who decide to express their pay in lieu of giving sufficient documentation. Further, the strategy and framework can be material to any retail loaning business including, however not restricted to, contract, vehicle advance, and Mastercards, where credit authority data shapes a piece of the information assortment process and is accessible alongside candidate's data. The past area utilizes a graphical way to deal with assess the exhibition of our straight model, which can in some cases be very abstract in our judgment. Here we will utilize a few quantitative strategies to acquire a more exact presentation assessment of our straight model.

- According to claim1# the invention is to a "Compare the Dates in a Month with the Monthly Salary using Regression Method" is a breaks down the connection between private pay and the typical number of hours worked each week. Work market financial matters infer a positive connection between hours worked and higher pay, yet in addition recommend that beyond a particular pay edge, people will more often than not work less hours.

- According to claim1,2 # the invention is to a connection between hours worked and individual pay is demonstrated utilizing information from the United States Census Bureau's American Community Survey from 2013 and 2019. Straightforward relapse investigation tried the log of changed individual pay against hours worked, and the different relapse extended this examination to incorporate gas utility costs, number of laborers in family, food stamp help, and number of people in family as factors.
- According to claim1,2,3# the invention is to a basic relapse model exhibits that individual pay has a positive relationship with hours worked, while the numerous relapse model shows that this impact reduces as pay level increments. A mechanized strategy and framework gauges pay of a singular advance candidate utilizing credit authority data and advance ascribes.
- According to claim1,2,3,4# the invention is to a strategy and framework can utilize the credit department and advance data to adjust a candidate's obligation trouble in situations where such data is not promptly accessible or is mysterious. The strategy and framework can naturally check pay for candidates who decide to express their pay in lieu of giving sufficient documentation.
- According to claim1,2,3,4,5# the invention is to a strategy and framework can be material to any retail loaning business including, however not restricted to, contract, vehicle advance, and Mastercards, where credit authority data shapes a piece of the information assortment process and is accessible alongside candidate's data.
- According to claim1,2,3,4,6# the invention is to a past area utilizes a graphical way to deal with assess the exhibition of our straight model, which can in some cases be very abstract in our judgment. Here we will utilize a few quantitative strategies to acquire a more exact presentation assessment of our straight model.

REFERENCES

[1] https://www.grammarly.com/?q=researchpaper&utm_source=bing&utm_medium=cpc&utm_campaign=420521273&utm_content=&utm_term=%2Bresearch%20%2Bpaper&matchtype=p&placement=&network=s&&msclkid=b7411d67a8e9199930147f16d896af8a&gclid=b7411d67a8e9199930147f16d896af8a&gclsrc=3p.ds

[2] https://www.ijsr.net/?msclkid=03ea0427bfe81731bbe86656cf9763c6

[3] mazon.in/s/?ie=UTF8&keywords=paper+paper&tag=msndeskstdin-21&index=aps&hvadid=75591586504176&hvqmt=p&hvbmt=bp&hvdev=c&ref=pd_sl_37worj3hrf_p&tag=msndeskstdin-21&ref=pd_sl_37worj3hrf_p&adgrpid=1209463

[4] https://dopt.gov.in/monthly-remuneration

[5] https://doe.gov.in/brochure-on-pay-and-allowances%20archives

[6] https://cleartax.in/s/7th-pay-commission-latest-news

CITATIONS

- Jain, A., Kumar, A. and Sharma, S., 2015. Comparative Design and Analysis of Mesh, Torus and Ring NoC. Procedia Computer Science, 48, pp.330-337.
- Jain, A., Dwivedi, R., Kumar, A. and Sharma, S., 2017. Scalable design and synthesis of 3D mesh network on chip. In Proceeding of International Conference on Intelligent Communication, Control and Devices (pp. 661-666). Springer, Singapore.
- Jain, A., Gahlot, A.K., Dwivedi, R., Kumar, A. and Sharma, S.K., 2018. Fat Tree NoC Design and Synthesis. In Intelligent Communication, Control and Devices (pp. 1749-1756). Springer, Singapore.
- Gupta, N., Jain, A., Vaisla, K.S., Kumar, A. and Kumar, R., 2021. Performance analysis of DSDV and OLSR wireless sensor network routing protocols using FPGA hardware and machine learning. Multimedia Tools and Applications, 80(14), pp.22301-22319.

Chapter - 13

Cost Effective Vehicle Monitoring on Construction Sites using AI

Ram Singh, Karan Babbar, Himanshu Kandpal, Hari Krishna

Quantum University, Dehradun Highway, Mandawar, Roorkee, Uttarakhand 247167, India.

ABSTRACT

Our Invention "Cost Effective Vehicle Monitoring System On Construction Sites and its Applications" The arrangement of all featured issues above in the issue area can be settled by utilizing just an upgraded vehicle observing framework for certain incredible changes and new innovation which will suit building site conditions and will facilitate the tension on Construction/Research Managers or can say of an association. While vehicle identification on interstates has been accounted for previously, apparently, clever observing framework that targets recognizing water powered backhoes and dump trucks on state-claimed land has not been investigated completely yet. In this invention, we present a programmed, video-based calculation for identifying water driven backhoes and dump trucks. Gotten from illustrations gained from video handling, we proposed strategies for frontal area recognition dependent on a further developed edge contrast calculation, and afterward recognized water driven tractors and dump trucks in the individual locale of interest. From our examination, we proposed strategies dependent on reverse valley element of mechanical arm and spatial-worldly thinking for water driven earthmover identification. Furthermore, we investigated dump truck recognition procedures that consolidate organized part research ion with the spatial relationship. Investigates genuine checking destinations showed the promising exhibition of our framework.

FIELD OF THE INVENTION

Our Invention is related to a Cost Effective Vehicle Monitoring System On Construction Sites and its Applications.

BACKGROUND OF THE INVENTION

Assurance of state-possessed land is a significant obligation for each neighborhood land and assets department because of the quick urbanization in China. Indeed, even with such severe management, state-claimed land is infringed upon consistently by engineers and people.

In 2010, 372 regular unlawful occupation cases happened in Shenzhen, the primary extraordinary monetary zone in China. Most of these cases are illicit development on state-claimed land.

To admonish illicit development without really wasting any time, a clever checking framework pointed toward identifying development vehicles on locales is presented. An assortment of designing vehicles are utilized in weighty common development. Large numbers of these vehicles are made to do explicit tasks, while others, like water driven tractors and dump.

Utilizing an insightful checking framework to admonish of unlawful development can significantly save money on cost. When the observing framework is constructed, all handling systems are completely programmed. The framework cannot just give total data of each video site, yet in addition nitty gritty data of each pre put position.

Moreover, the framework we propose has great ongoing execution and high productivity; it can deal with recordings progressively and handle different recordings all the while.

Our framework is additionally completely programmed, and subsequently no particular oversight is required. As per our examination, studies or calculations identified with the discovery of water driven earthmovers and dump trucks on state-claimed land don't exist.

In this manner, assets that can be learned and utilized for reference is restricted. Through our examination and on-the-spot examination, we accumulated that a cautioning before illicit development for multiple layers is fundamental.

Based on this examination, admonishing data ought to be given before development, that is, during the stage when establishments are being laid or development exercises have recently begun. This examination agrees with our recognition target in light of the fact that pressure driven earthmovers can be utilized in the establishment laying stage, which for this situation incorporates unearthing, stacking, managing, and moving materials, just as in the development stage. Moreover, dump trucks can be utilized in the establishment laying stage for grime transportation and in the development stage for building material transportation.

PRIOR ART SEARCH

Building locales are dynamic and convoluted frameworks. The development and cooperation of individuals, products and energy make development security the executives amazingly troublesome. Due to the consistently expanding measure of data, conventional development security the board has worked under troublesome conditions.

As a compelling method for gathering, distinguish and process data, sensor-based innovation is considered to give new age of techniques to propelling development security the board. It makes the continuous development wellbeing the board with high effectiveness and precision a reality and gives a strong establishment to working with its modernization, and Informatization.

These days, different sensor-based advancements have been taken on for development security the executives, including finding sensor-based innovation, vision-based detecting and remote sensor organizations. This invention gives an efficient and exhaustive audit of past investigations in this field to recognize valuable discoveries, distinguish the exploration holes and point out future examination bearings.

OBJECTIVES OF THE INVENTION

The objective of the invention is to provide "Cost Effective Vehicle Monitoring System On Construction Sites and its Applications" The arrangement of all featured issues above in the issue area can be settled by utilizing just an upgraded vehicle observing framework for certain incredible changes and new innovation which will suit building site conditions and will facilitate the tension on Construction/Research Managers or can say of an association.

The other objective of the invention is to provide a vehicle identification on interstates has been accounted for previously, apparently, clever observing framework that targets recognizing water powered backhoes and dump trucks on state-claimed land has not been investigated completely yet.

The other objective of the invention is to provide a programmed, video-based calculation for identifying water driven backhoes and dump trucks. Gotten from illustrations gained from video handling, we proposed strategies for frontal area recognition dependent on a further developed edge contrast calculation, and afterward recognized water driven tractors and dump trucks in the individual locale of interest.

The other objective of the invention is to provide an examination, we proposed strategies dependent on reverse valley element of mechanical arm and spatial-worldly thinking for water driven earthmover identification.

The other objective of the invention is to provide Furthermore, we investigated dump truck recognition procedures that consolidate organized part research ion with the spatial relationship. Investigates genuine checking destinations showed the promising exhibition of our framework.

SUMMARY OF THE INVENTION

The invention embraces a systemic way to deal with lead a far reaching survey of sensor-based innovation applications in development security the executives. It gives an establishment to investigating helpful discoveries and recognizing holes for future exploration based on past examinations.

With the assistance of logical choice strategies, key inventions applicable to the picked theme were chosen and gathered to develop an information base. Information examination was then performed from sequential and topical points of view to distinguish helpful exploration discoveries.

A two-stage writing choice strategy after Tsai and Wen was applied to gather key important inventions. The time not set in stone to be from 2005 to 2016. The explanation is that before 2005, barely any sensor-based advances were used in development wellbeing the executives and their application range was very limited. In the interim, the degree of innovation during this period was excessively straightforward, and the substance of significant inventions would not assume a positive part in further developing development wellbeing the board.

Concerning the determination of sensor-based innovation, it is led in the accompanying advances:

- in view of the survey of important inventions, some sensor-based advances were for starters distinguished since their use was relevant to further develop development security the board;
- for those advances with little writing accessible (just 1 or 2 references) that just expressed their ability of further developing development wellbeing the executives without check and test measures (e.g., laser output and infrared),

There are some fundamental issues which we face on building destinations in everyday exercises. Here are some featured list items of issues recorded underneath:

1. Thefts.
2. Supervise driver conduct.
3. Getting dependable information of vehicles during activity.
4. Location following of Vehicles on location.
5. Fuel Monitoring.
6. Accidents.

A Survey was Conducted in 2020 which brought about an immense number of issues looked on the site for not embracing the innovation:

Development associations are mulling over place of work worker following in 57% of cases. Absence of IT experts was shown by 38.2 percent of development associations as the primary justification for not executing new innovation for research the board.

Practically 90% of overall foundation researchs are postponed or ended because of a lack of hardware or a spending plan that is excessively high. For a given video arrangement, the frontal area discovery calculation, which is the further developed three-outline differential technique, is performed first to decide whether an up-and-comer dump truck is available.

Then, at that point, we see as the "V component" in the competitor wheel region and decide the ROI of the taxi and container as indicated by the underlying relationship of the landfill truck. For taxi and container recognition, we utilize a technique that settles on a choice dependent on researchion.

In examinations and overviews, various researchers are dedicated to vehicle identification, particularly, vehicles on thruway that present the front or back view. Writing which about dump trucks that present the side-view is restricted. Given the viable utilization of our undertaking, cameras are held tight structures or in the corner.

BRIEF DESCRIPTION OF THE DIAGRAM

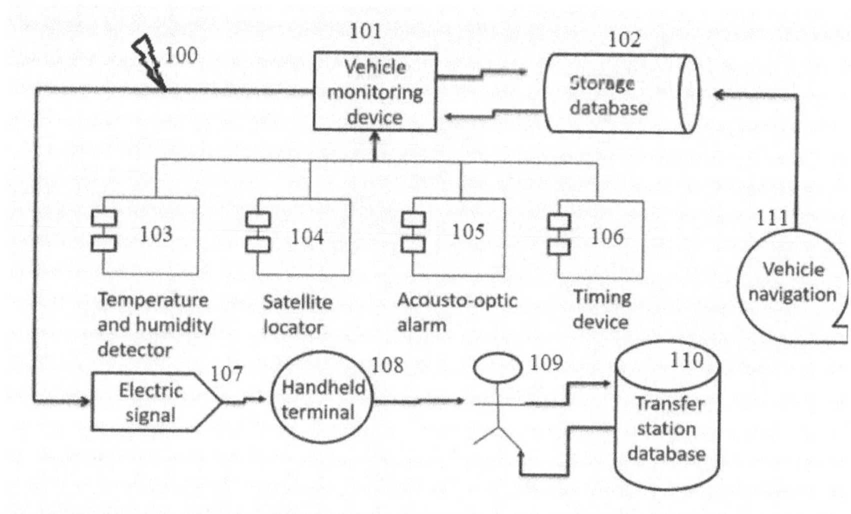

Fig. 1: Cost Effective Vehicle Monitoring System On Construction Sites, Process.

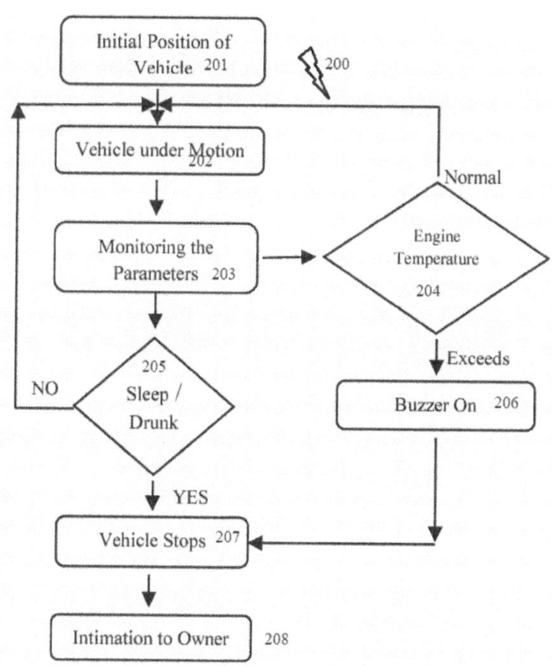

Fig. 2: Cost Effective Vehicle Monitoring System On Construction Sites Process.

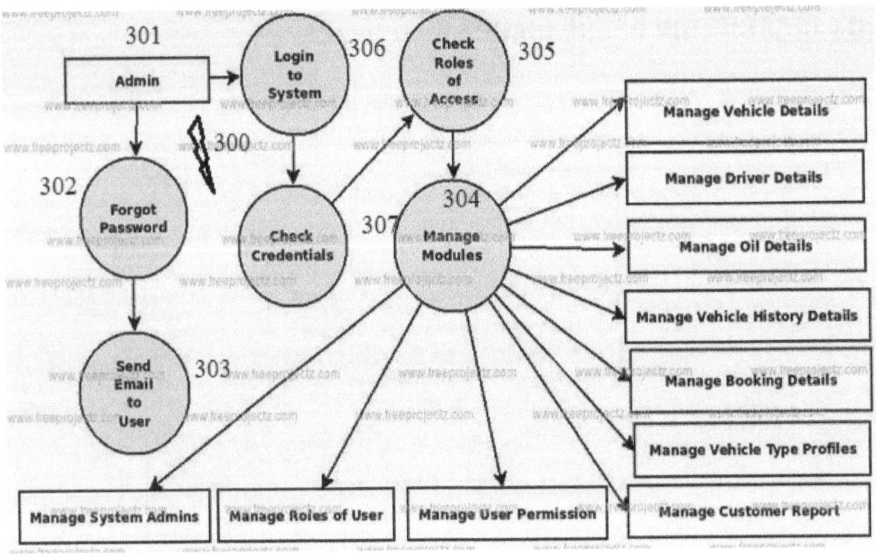

Fig. 3: Cost Effective Vehicle Monitoring System On Construction Sites and its Applications.

DESCRIPTION OF THE INVENTION

- The vehicle checking framework will dispose of straight forwardly the issues of burglary on the site.
- We will get solid information right on our cell phones or PCs
- The cash which will be saved by the organization/association will be enormous.
- The framework will screen driver's conduct which will give us the information immediately that driver driving status like high speed increase, over speeding, lazy driving which will assist us with expanding the strength of a development vehicle to utilize it for a more drawn out term.
- As we will screen driver's conduct we can diminish mishaps on the locales in light of driver's slip-ups.
- Area Tracking of the development vehicles should be possible with this framework.
- This framework will likewise influence by implication outside climate then the building locales.

The Market for the Product: Essentially, this item will focus available of substantial development vehicles where the requirement for this innovation will tackle there some significant issues identified with their everyday exercises and will save them huge load of cash which will draw in the interest for this innovation. Furthermore, this item will cost-be accommodating to every one of the purchasers which will likewise diminish the utilization of labor nearby.

This Technology can be utilized with no problems in the two Governments like PWD, CIDCO, MHADA, and a lot greater Government areas additionally in Private Sectors like GVK, L&T, TATA and a lot more such associations.

Water driven backhoe is a sort of water driven gear, that is chiefly utilized for removal. It is broadly utilized in stacking, managing, and moving materials, and is, accordingly, generally utilized in development exercises. Through perception, we observed that the development of a pressure driven earthmover is generally lethargic contrasted and different vehicles.

Consequently, utilizing ceaseless two edges to decide the thing that matters is lacking. Through investigation and examinations, the recognized exhibition of utilizing the two casings a good way off of 4 is accomplished.

Edge detection: Edge identification is performed. To recognize Inverse-V element of the mechanical arm of water powered earthmover, slanting Sobel cover for edge identification is utilized, which thinks about the force reaction of the inclining edge. 2) Binary activity utilizing the possibility of division: After the dim scale picture which demonstrates edge reaction is acquired, the subsequent stage is to perform edge activity normally. Worldwide and fractional edges are two principle techniques utilized for limit activity.

Nonetheless, a powerful technique that can smother nearby commotion as well as keep solid edges is expected to get the paired picture. Enlivened by picture division, dark scale picture can be separated into two classes, specifically, solid edges and foundation and commotion. Along these lines, the fundamental issue is to portion the two classes. We can utilize the strategy referenced in Section 2 to portion the picture.

Contour filter: Through the above interaction, a double picture that addresses edges is acquired. Be that as it may, some irritating commotions and pseudo edges might in any case exist. Morphological activities are ordinarily utilized for additional handling; however, such activities are probably not going to satisfy this assignment due to foundation intricacy.

Taking into account that the mechanical arm shape is bigger than the rest and that other dissipated forms are moderately little, we use shape channel to refine the double picture. In our venture, we dispose of the forms whose limiting square shapes are more noteworthy than 256*256 or under 64*64, in light of the fact that it can eliminate little, irritating layouts yet in addition crash enormous forms, which are brought about by missing information.

Line discovery: As expressed in Chapter 2, the containers, in which the lines of blast and scoop have a place with ought to be recognized. It is helpful to affirm line points that address the blast and scoop.

RESEARCH CLAIM

Our Invention "Cost Effective Vehicle Monitoring System On Construction Sites and its Applications" The arrangement of all featured issues above in the issue area can be settled by utilizing just an upgraded vehicle observing framework for certain incredible changes and new innovation which will suit building site conditions and will facilitate the tension on Construction/Research Managers or can say of an association. While vehicle identification on interstates has been accounted for previously, apparently, clever observing framework that targets recognizing water powered backhoes and dump trucks on state-claimed land has not been investigated completely yet. In this invention, we present a programmed, video-based calculation for identifying water driven backhoes and dump trucks. Gotten from illustrations gained from video handling, we proposed strategies for frontal area recognition dependent on a further developed edge contrast calculation, and afterward recognized water driven tractors and dump trucks in the individual locale of interest. From our examination, we proposed strategies dependent on reverse valley element of mechanical arm and spatial-worldly thinking for water driven earthmover identification. Furthermore, we investigated dump truck recognition procedures that consolidate organized part research ion with the spatial relationship. Investigates genuine checking destinations showed the promising exhibition of our framework.

- According to claim1# the invention is to provide a "Cost Effective Vehicle Monitoring System On Construction Sites and its Applications" The arrangement of all featured issues above in the issue area can be settled by utilizing just an upgraded vehicle observing framework for certain incredible changes and new innovation which will suit building site conditions and will facilitate the tension on Construction/Research Managers or can say of an association.

- According to claim1,2# the invention is to provide a vehicle identification on interstates has been accounted for previously, apparently, clever observing framework that targets recognizing water powered backhoes and dump trucks on state-claimed land has not been investigated completely yet.
- According to claim1,2,3# the invention is to provide a programmed, video-based calculation for identifying water driven backhoes and dump trucks. Gotten from illustrations gained from video handling, we proposed strategies for frontal area recognition dependent on a further developed edge contrast calculation, and afterward recognized water driven tractors and dump trucks in the individual locale of interest.
- According to claim1,2,3,4# the invention is to provide an examination, we proposed strategies dependent on reverse valley element of mechanical arm and spatial-worldly thinking for water driven earthmover identification. Furthermore, we investigated dump truck recognition procedures that consolidate organized part research ion with the spatial relationship. Investigates genuine checking destinations showed the promising exhibition of our framework.

REFERENCES

[1] Arabi, S., Haghighat, A., and Sharma, A. (2020). "A deep-learning-based computer vision solution for construction vehicle detection." Computer-Aided Civil and Infrastructure Engineering, 1–15. 52 Bugler,

[2] M., Borrmann, A., Ogun Makin, G., Vela, P. A., and Teaser, J. (2017). "Fusion of Photogrammetry and Video Analysis for Productivity Assessment of Earthwork Processes." Computer-Aided Civil and Infrastructure Engineering, 32(2), 107–123.

[3] Cheng, J. C. P., and Wang, M. (2018). "Automated detection of sewer pipe defects in closed-circuit television images using deep learning techniques." Automation in Construction, Elsevier, 95(August), 155–171. Darko, A., Chan, A. P. C., Adabre, M.

[4] A., Edwards, D. J., Hosseini, M. R., and Amaya, E. E. (2020). "Artificial intelligence in the AEC industry: Scientometrics analysis and visualization of research activities." Automation in Construction, Elsevier, 112(May 2019), 103081.

[5] Fang, Q., Li, H., Luo, X., Ding, L., Luo, H., and Li, C. (2018). "Computer vision aided inspection on falling prevention measures for steeplejacks in

an aerial environment." Automation in Construction, Elsevier, 93(December 2017), 148–164.

[6] Fang, W., Ding, L., Love, P. E. D., Luo, H., Li, H., Peña-Mora, F., Zhong, B., and Zhou, C. (2020a). "Computer vision applications in construction safety assurance." Automation in Construction, Elsevier, 110(February 2019), 103013. Fang, W., Love, P.

[7] E. D., Luo, H., and Ding, L. (2020b). "Computer vision for behaviour-based safety in construction: A review and future directions." Advanced Engineering Informatics, Elsevier, 43(August 2019), 100980.

[8] Fang, W., Zhong, B., Zhao, N., Love, P. E. D., Luo, H., Xue, J., and Xu, S. (2019). "A deep learning-based approach for mitigating falls from height with computer vision: Convolutional neural network." Advanced Engineering Informatics, Elsevier, 39(December 2018), 170–177.

[9] Golparvar-Fard, M., Peña-Mora, F., and Savarese, S. (2015). "Automated progress monitoring using unordered daily construction photographs and IFC-based building information models." Journal of Computing in Civil Engineering, 29(1), 1–19.

CITATIONS

- Jain, J., Jain, A., Srivastava, S. K., Verma, C., Raboaca, M. S., & Illés, Z. (2022). Improved Security of E-Healthcare Images Using Hybridized Robust Zero-Watermarking and Hyper-Chaotic System along with RSA. Mathematics, 10(7), 1071.
- Jain, J., & Jain, A. (2022). Securing E-Healthcare Images Using an Efficient Image Encryption Model. Scientific Programming, 2022.
- Kumar, S., Shailu, A., Jain, A., & Moparthi, N. R. (2022). Enhanced Method of Object Tracing Using Extended Kalman Filter via Binary Search Algorithm. Journal of Information Technology Management, 14(Security and Resource Management challenges for Internet of Things), 180-199.
- Jain, A., Dwivedi, R. K., Alshazly, H., Kumar, A., Bourouis, S., & Kaur, M. (2022). Design and Simulation of Ring Network-on-Chip for Different Configured Nodes. CMC-COMPUTERS MATERIALS & CONTINUA, 71(2), 4085-4100.
- Sharma, S.K., Jain, A., Gupta, K., Prasad, D. and Singh, V., 2019. An internal schematic view and simulation of major diagonal mesh network-on-chip. Journal of Computational and Theoretical Nanoscience, 16(10), pp.4412-4417.

CHAPTER - 14

Frequency Identification using IoT and AI

Ajay Kumar Sharma, Chiranjit Maji, Chirag Malik

Quantum University, Dehradun Highway, Mandawar, Roorkee, Uttarakhand 247167, India.

ABSTRACT

Alongside the expedient improvement of detecting frameworks, remote transmission innovation, and increasingly fast chip, the extent of utilization of the Internet of things (IoT) has considerably and quickly expanded, and examination and development that incorporate computerized reasoning. At same time, grounds application development as the groundwork of knowledge schooling is being on the walk, which requests recently arising PC advances support, like IoT, AI, Big information, etc. In this review, we give an engineering of brilliant grounds in view of Radio Frequency Identification (RFID) which can offer new answers for tackle a few issues like position, distinguishing proof, store data, etc as a result of its solid antiinterference capacity and non-meddling, as center innovation of IoT. Then in light of this design, a sort of savvy grounds application framework conspire is proposed, which can understand the capacity of programmed character acknowledgment, book perusing, inward utilization, understudy positions. Besides, for some, unique definite area requests, this paper proposes a strategy to definitively gauge RFID label area in view of triangulation standard, which is installed in the engineering. The utilization of Internet of things innovation can more readily advance some beginning or organization productivity, like more proficient assistance for educators' and understudies' work, study and life.

Keywords: *IoT; Smart campus; RFID; Sensor technology, Antiinterference, Capacity, Non-meddling.*

INTRODUCTION

After IBM proposed "brilliant planet" in 2018, savvy grounds in China is being built, fanning out from colleges and universities to optional schools, even grade schools. The quick improvement of current organization correspondence innovation and the fast advancement of PC related equipment hardware have advanced the speed of the development of University data, as of now, have assembled different scale data computerized grounds, which has understood the mix of data assets, electronic installment and monetary framework and other data the board frameworks partially, and has carried a ton of comfort to the work, study and life of educators and understudies.

For instance, the use of Campus IC card framework has given an advantageous way to understudies in school feasting, shopping, library, etc. In 1994, Schilit and Theimer originally proposed the idea of the setting mindfulness.

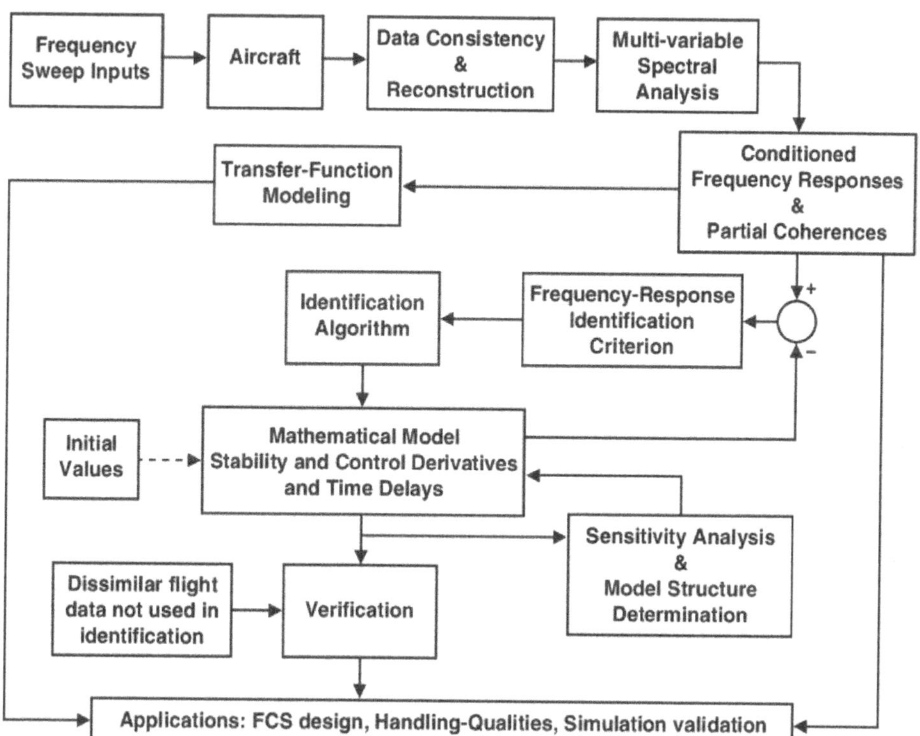

Fig. 1: Frequency Identification using IoT and AI

With the extending of exploration, Dey presented the idea of versatile (Dey, et al.1998) [3-4]. Salber presented another definition (Salber, et al. 1998) [5]. Dey and Abowd further extended the field of utilization of setting mindfulness (Dey, et al. 1999) [6]. Tewari fostered an extensive shrewd choice emotionally supportive network.

Definition of the Internet of Things

The IoT is a significant piece of the new age of data innovation. Single word in its name is "Internet+", so the Internet of things can be perceived as "the material articles associated with the Internet. Profoundly and the groundwork of the Internet of things is as yet the Internet, it is the augmentation and development of the organization in view of the web; Secondly, its client side stretches out and grows to any things to carry on the data trade and the correspondence.

Web of things is the mix application in view of the insightful detecting, acknowledgment innovation and unavoidable processing, universal organization, it is known as the world's third flood of data industry advancement following the PC and the Internet. Web of things is the application advancement of the Internet. Profoundly.

Connectivity

Availability is the premise of the Internet of things. Network, remote, wired or detecting objects should mirror the "association" state. The International Telecommunications Union accepts that the availability of the Internet of things incorporates four viewpoints: the first is that it can carry on the association whenever: the second is that it can carry on the association in any spot; the three is that it very well may be associated with any item, the four is that it tends to be associated with any individual.

Physical Object Association

In the Internet, individuals use PCs to speak with one another, while the Internet of things through the sensor, radio recurrence recognizable proof innovation, worldwide situating framework and other innovation to accomplish the trading of things and articles. Web accomplishes the significant distance correspondence between the endlessly individuals, and the Internet of things accomplishes a continuous communication among individuals and items, articles and things, to accomplish the progress from the virtual world to this present reality.

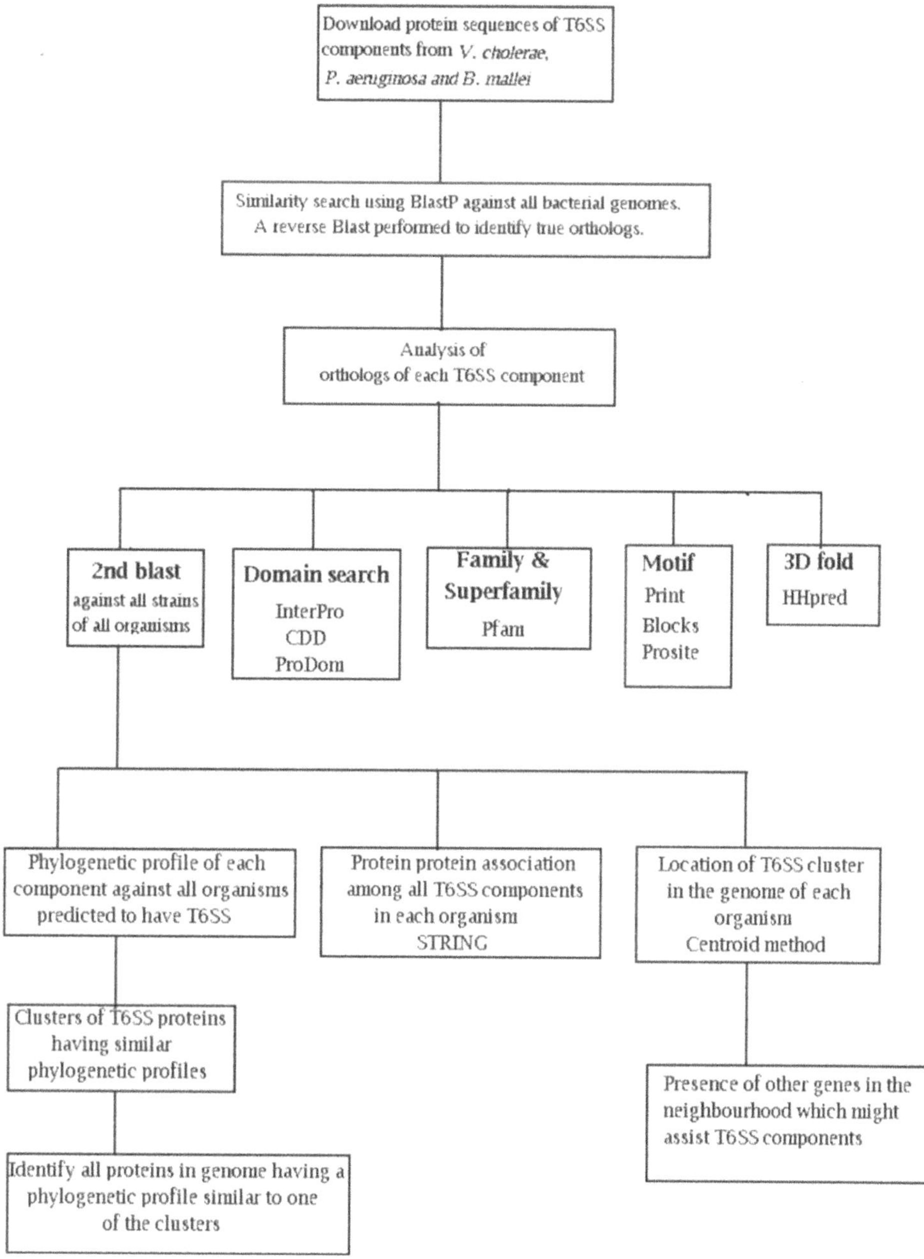

Fig. 2: Frequency Identification using IoT and AI, Block Diagram

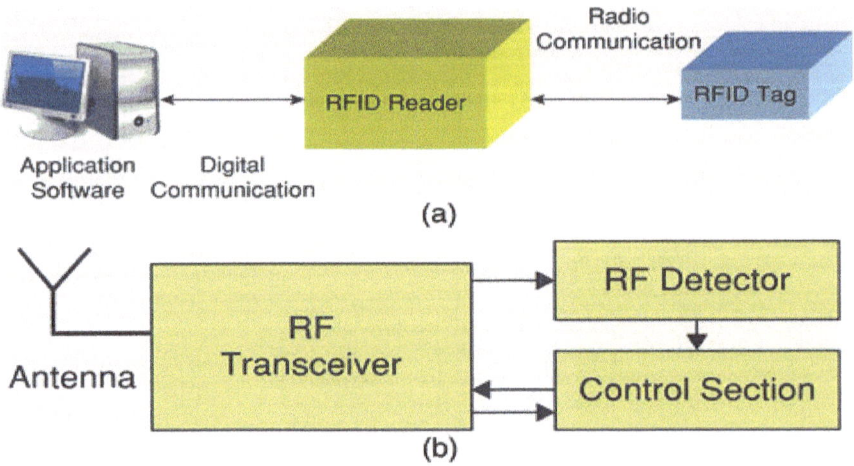

Fig. 3: Frequency Identification using IoT and AI Process

Intelligence

Intelligence alludes to complete the smart control for each article by the utilization of PC innovation, sensor innovation and control innovation and wise checking framework, starting here of view, the Internet is a significant piece of the Internet of things, which makes that the items on the planet not exclusively can be related by the approach to detecting, however the method of canny.

Web of things has the smart discernment, it can detect the climate, and augment the help of individuals to better understanding into the utilization of an assortment of assets, to make a more precise judgment. In this way, the Internet of things has been broadly utilized in different fields, and has created the business chain. As of now, a few schools and colleges in China have set up the Internet of things, and have explored the insightful sensor organization, which has established the groundwork for the further advancement of the Internet of things innovation.

Requirement

Examination of the Construction of Smart Campus Internet of things industry chain can be partitioned into: distinguishing proof, discernment, data handling and transmission of 4 connections, the vital innovations of each connection are RFID, sensor, brilliant chip and the remote transmission organization of the remote worker administrators.

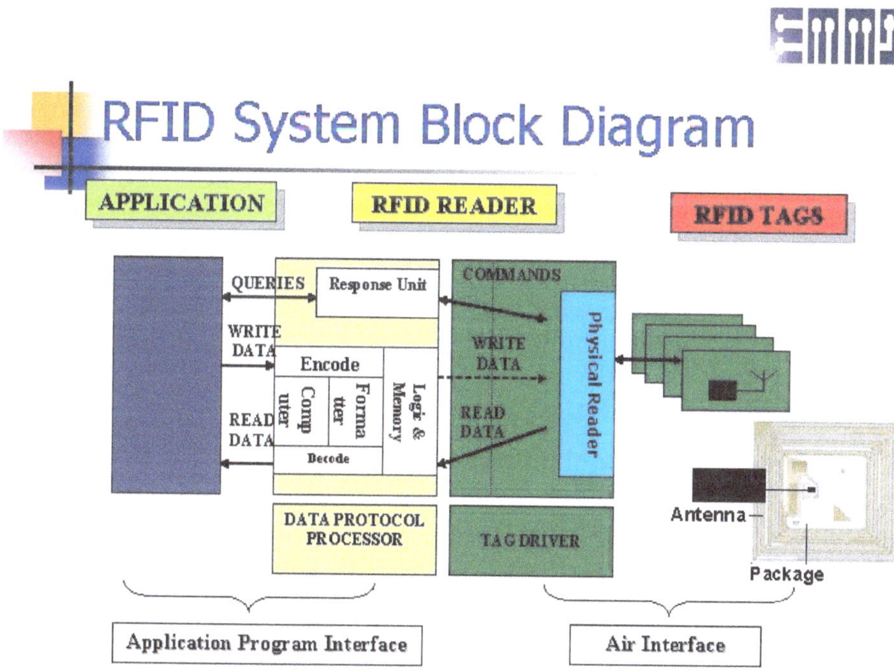

Fig. 4: Frequency Identification using IoT and AI date proc.

Considering the issues that the homegrown Internet of things industry establishment and the key connection were somewhat powerless, the normalization work was moderately in reverse, etc, the public association began the Internet of things innovation innovative work and industrialization projects, sent off a public systems administration application showing project, wanted to advance the turn of events and industrialization of key innovations and modern, standard framework and public assistance stage, and zeroed in on the forward leaps in center advancements, and the improvement of the modern chain, in order to offer a successful help for the utilization of key region of the Internet of things.

Methodology

According to the basic idea of the effective integration of various functions of digital campus and the user's convenience, based on the Internet of things technology, this paper presents a design scheme of intelligent campus management system, which can realize the functions of user management and automatic identification, library borrowing, internal consumption, student information management and so on.

Architecture

According to the demand of the automatic identification, application data management integration and sharing of teachers and students in the era of Internet of things intelligent digital campus, the intelligent digital campus based on Internet of things adopts the three-tier system, as shown in Figure 2, which mainly includes the application data service layer, network communication layer and terminal user layer.

System Function Module

Tag reader Smart mobile phone Based on the unified user identity recognition mechanism, the intelligent digital campus system based on Internet of things mainly consists of user identity management, conference management, student attendance management, school vehicle management of four function modules. User identity management module: it through the sensing technology and RFID technology to achieve personnel in and out of control in library and other important places, in addition, it also comes with recording function, which can count the passengers and time through the signal.

School vehicle management module: it uses RFID technology to achieve rapid, real-time, accurate acquisition and processing for the information, besides, it is regarded as the upgrade and supplement of the traditional vehicle management methods to meet the requirements of intelligent campus vehicle management.

CONCLUSION

The smart campus achieves intelligent campus services and management through the Internet of things to, it can achieve the interconnection of any person, anything, any information carrier, any time, any place through the Internet of things, besides, the polymerization of the vast amounts of information in the Internet of things platform provides smart business and service model for the majority of teachers and students.

There is no doubt that the smart campus has brought great convenience to teachers and students, but also has faced many challenges. Firstly, there is the issue of cost, the construction of the smart campus must have a unified infrastructure platform, including wired and wireless dual network coverage of the network environment.

REFERENCES

[1] Geng Zhen J 2022 A new stage of educational Informatization: from "digital campus" to "smart compus" (Computer Era) p 76-79+83

[2] Schiliti J 2020 Disseminating active map information to mobile host vol 5(IEEE Network) p 2232

[3] Dey A.K C 2019Context-Aware Computing: The Cyber Desk Project In AAAI (Spring Symposium on intelligent environments technical report) p51-54.

[4] Dey K. Anind and Abowd C 2019 Gregory towards Better Understanding of Context and Context Awareness Technical report (GIT-GVU-99-22 Georgia Institute of Technology)

[5] Salber D. Dey A.K. and Abowd C 2018 Ubiquitous computing: defining and HCI research agenda for an emerging paradigm (Georgia Tech GVU Technical Report GIT-GVU-98-01)

[6] Dey A.K J 2019 Cyber Desk: A framework for providing self-integrating context-aware services vol 1 ed Abowd GD, Wood A (Knowledge-Based Systems) p3-13

[7] Tewari J 2018 Personalized location-based brokering using an agent-based intermediary architecture vol 34 ed G; Youll J; Maes, P (Decision support systems) p127-137

[8] Tseng KuoHsiung J 2021 Green Smart Campus Monitoring and Detection Using LoRa vol 21 ed Chung MengYun, Chen LiHsien, Chang PeiYao (Sensors) p6582-6582

[9] Ma D H J 2011 Internet of Things: Objectives and Scientific Challenges vol 11 (Journal of Computer Science & Technology) p15.

CITATIONS

- Jain, A., Dwivedi, R. K., Alshazly, H., Kumar, A., Bourouis, S., & Kaur, M. (2022). Design and Simulation of Ring Network-on-Chip for Different Configured Nodes. CMC-COMPUTERS MATERIALS & CONTINUA, 71(2), 4085-4100.
- Jain, A., Kumar, A. and Sharma, S., 2015. Comparative Design and Analysis of Mesh, Torus and Ring NoC. Procedia Computer Science, 48, pp.330-337.
- Jain, A., Dwivedi, R., Kumar, A. and Sharma, S., 2017. Scalable design and synthesis of 3D mesh network on chip. In Proceeding of International Conference on Intelligent Communication, Control and Devices (pp. 661-666). Springer, Singapore.
- Jain, A., Gahlot, A.K., Dwivedi, R., Kumar, A. and Sharma, S.K., 2018. Fat Tree NoC Design and Synthesis. In Intelligent Communication, Control and Devices (pp. 1749-1756). Springer, Singapore.
- Jain, A., Dwivedi, R., Kumar, A. and Sharma, S., 2017. Scalable design and synthesis of 3D mesh network on chip. In Proceeding of International Conference on Intelligent Communication, Control and Devices (pp. 661-666). Springer.

Chapter - 15

IoT-based Raise Farming

Saloni Arora, Rajwinder Kaur, Praveen Kumar

Quantum University, Dehradun Highway, Mandawar, Roorkee, Uttarakhand 247167, India

ABSTRACT

Regardless of incredible triumphs in crop research for the creating scene, numerous unfortunate ranchers in peripheral regions have not benefited. There is expanding acknowledgment that harvest examination would be able and ought to help these ranchers. This paper audits a portion of the key specialized issues connected with the test of figuring out and completing yield research that is designated, applicable and suitable to these ranchers, with a specific accentuation on crop improvement. It inspects the reasons that support the absence of examination effects, and a portion of the critical difficulties and instruments accessible to get this going. The difficulties incorporate reproducing for minimal conditions, consolidating hazard and weakness as significant elements of harvest research, perceiving the interest for numerous attributes and different assortments, as well as addressing disregarded and underutilized species essential to poor people. The devices are contained a blend of yield ecological grouping frameworks with destitution planning to target and focus on research at the worldwide and public levels, the utilization of participatory exploration draws near, especially in plant rearing, as well as non-market valuation philosophies to make research items applicable and fitting to the poor at the neighborhood level. The paper momentarily inspects the guarantee of current biotechnology.

Keywords: *Targeting, Priorities, Poor, Marginal environments, Participatory breeding, Poverty mapping, Biotechnology, Non-market valuation.*

INTRODUCTION

Crop research, in plant reproducing, agronomy, crop security, post gather and handling along with related research in financial matters and other sociologies, has produced tremendous advantages for ranchers and customers in many pieces of the creating scene [1, 2]. In any case, countless ranchers, in various areas all through the creating scene, have been forgotten about. There is expanding acknowledgment that harvest examination would be able and ought to help these ranchers, and truth be told there are now different endeavors focused on this The focal point of yield research on preferred conditions has been supported because efficiency in them is high and hence profits from crop research speculation are high too [3].

Thusly it could be wasteful to designate greater venture away from more preferred creation regions to negligible regions where rural efficiency will in general be low [4]. An audit of the discussion on the venture on farming examination in preferred versus minimal conditions and its effect on destitution presumes that a large portion of the proof backings keeping a solid exploration accentuation on preferred creation conditions. Nonetheless, it additionally perceives that agrarian examination that objectives explicit troublesome creation conditions might be the most supportive of unfortunate public venture for a few negligible conditions [4].

In spite of the fact that accomplishing efficiency expansions in minor regions might be all the more expensive (in light of the fact that it is more hard to deal with these conditions) and the outright gains might be a lot of lower than in preferred regions, it has been contended that a 'the same old thing' way to deal with rural exploration might be shifty and untrustworthy [5]. Besides, there may not exactly be a compromise since there is proof that, inexorably, it is in minimal regions where ventures might give better yields because of consistent losses in preferred regions, especially on destitution decrease and ecological assurance.

Crop Research

Not Benefited Many of the Poor Farmers in the Developing World? There are many motivations behind why yield research has not helped numerous ranchers in the creating scene. The principal reason is that these ranchers are generally situated in troublesome conditions. The regions where they ranch are portrayed by profoundly factor and additionally outrageous precipitation conditions, troublesome geology, heterogeneous and normally unfortunate soils, as well as an absence of foundation, like streets, markets, water system, and public administrations, due to their area in difficult to-arrive at regions.

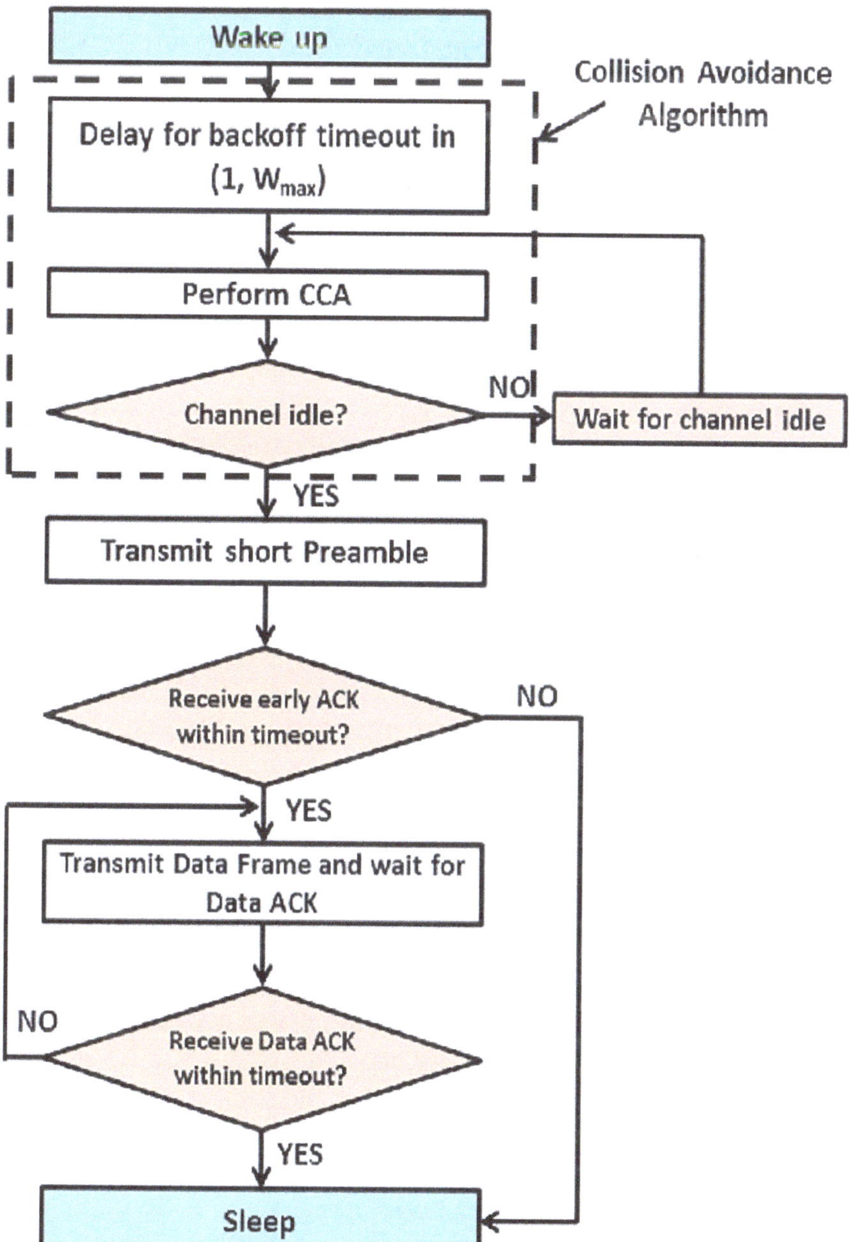

Fig. 1: IOT-based Raise Farming, FLow

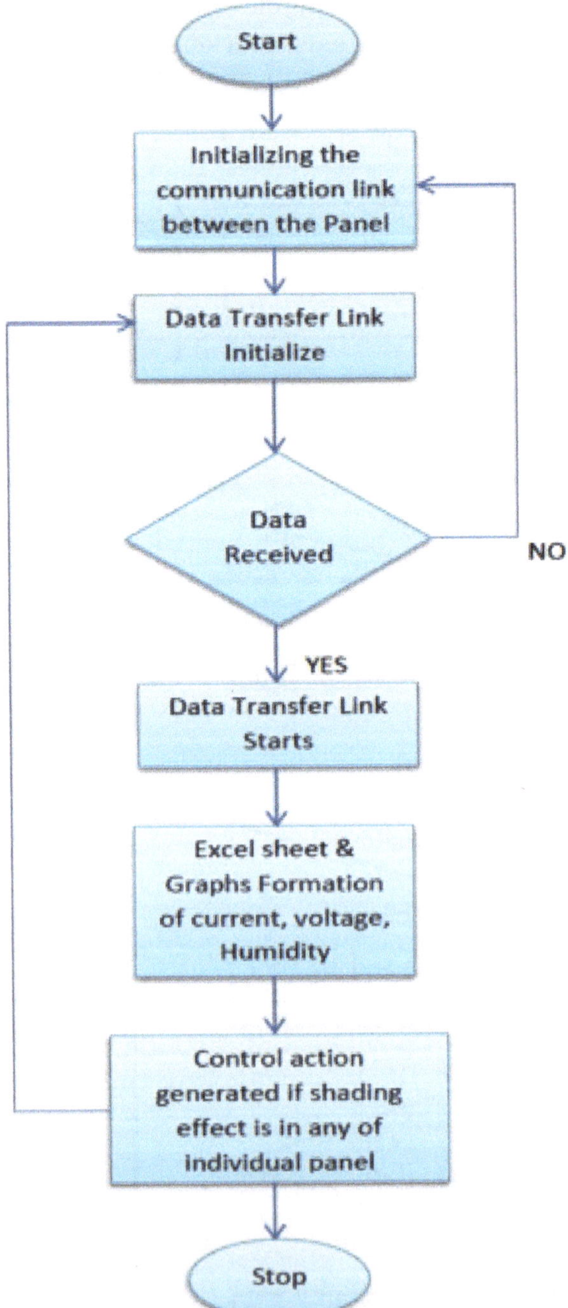

Fig. 2: IOT-based Raise Farming Flow Chart.

Besides, a large number of these ranchers are incorporated in native or socially minimized bunches that generally have not had a voice in their public or even territorial social orders, with somewhat minimal political power and capacity to impact government approaches and venture, and additionally inadequate with regards to the pay to draw in private revenue and speculation. Thirdly, a large number of these populaces are scattered, and keeping in mind that their populace thickness vis-a'- vis their asset base might be high (prompting solid tensions on it), their geological thickness might be low, making them more troublesome and exorbitant to reach per capita.

Incorporating Risk and Vulnerability

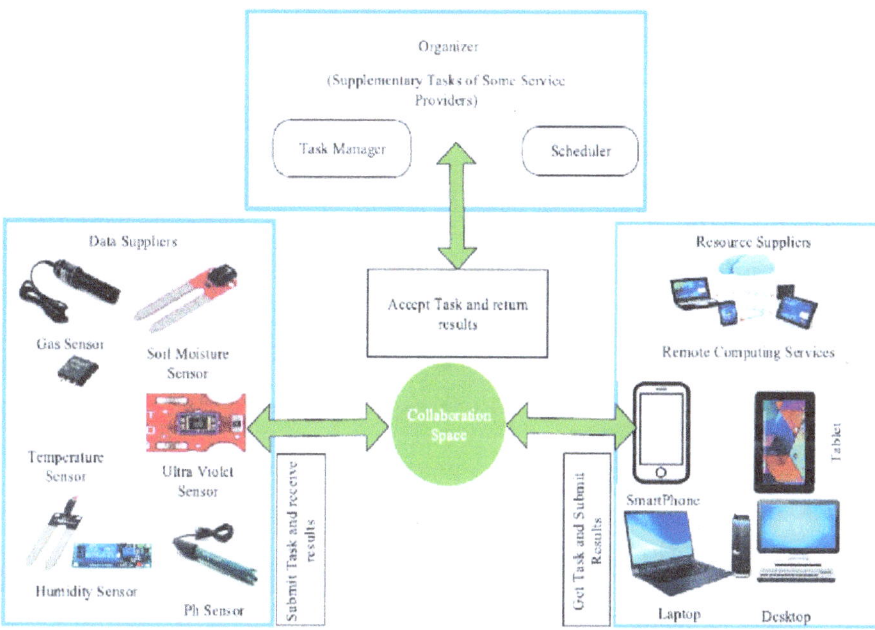

Fig. 3: IOT-based Raise Farming, Block Diagram

As Important Dimensions of Crop Research Farmers in minor regions face many dangers related with the conditions wherein they homestead and they have restricted ability to oversee them [4] 1] [5]. These dangers are not restricted to regular risks like dry season, irritation or infection, yet can likewise be related with stockpiling as well as showcasing of their items. Significant and suitable harvest research for these ranchers needs to think about those dangers where it can have an effect (for example dry season, housing and stockpiling vermin), straightforwardly delivering results that address them (for example delivering assortments that perform well under antagonistic ecological circumstances.

Multiple Crop

Qualities past Yield One of the main points of harvest research, especially crop reproducing, has been to build efficiency of different significant harvests. Efficiency is principally as far as expanding yield per unit of land as well as work, however progressively there is likewise interest in per unit of water.

Specific accentuation has been placed on expanding the yield potential from one viewpoint, and on decreasing the yield hole (the yield hole is typically characterized as contrast between normal yields in ranchers' fields and those of the most significant returns in research stations [1][8] while yield potential is 'the yield of a cultivar when filled in conditions to which it is adjusted, with supplements and water non-restricting, and with bothers, sicknesses, weeds, dwelling.

Targeting Crop

Research Targeting is a significant worry for crop research. The requirement for focusing on germplasm is very much perceived among reproducers and apparatuses, for example, the portrayal of creation conditions in view of agroclimatic factors and utilizing site-comparability strategies and measurable techniques, have been created to help this cycle.

This has prompted the advancement of order frameworks that partition the creation conditions of a yield into homogenous regions, as a rule on a worldwide or provincial scale. They are utilized by crop raisers to oversee genotype-by-climate associations and extrapolate inside comparative agroclimatic zones.

High-goal neediness planning is done at the nation level, which for focusing on crop research at a public level ought to be fine. The issue emerges to target crop research at a provincial or worldwide scale, since destitution lines normally contrast between nations. In this way, the individuals who are named poor in one nation might be grouped diversely in another even with comparable pay or use levels.

This implies that comparative agro-biological conditions in numerous nations can show various degrees of neediness rate under tantamount pay or use levels. It is significant in this situation to remember that while crop natural information might be similar at a territorial and worldwide level, neediness occurrence information may not be. Different kinds of financial planning might be more important, for example, guides of appetite or hunger [5][4][5][5]; however the fundamental thought continues as before.

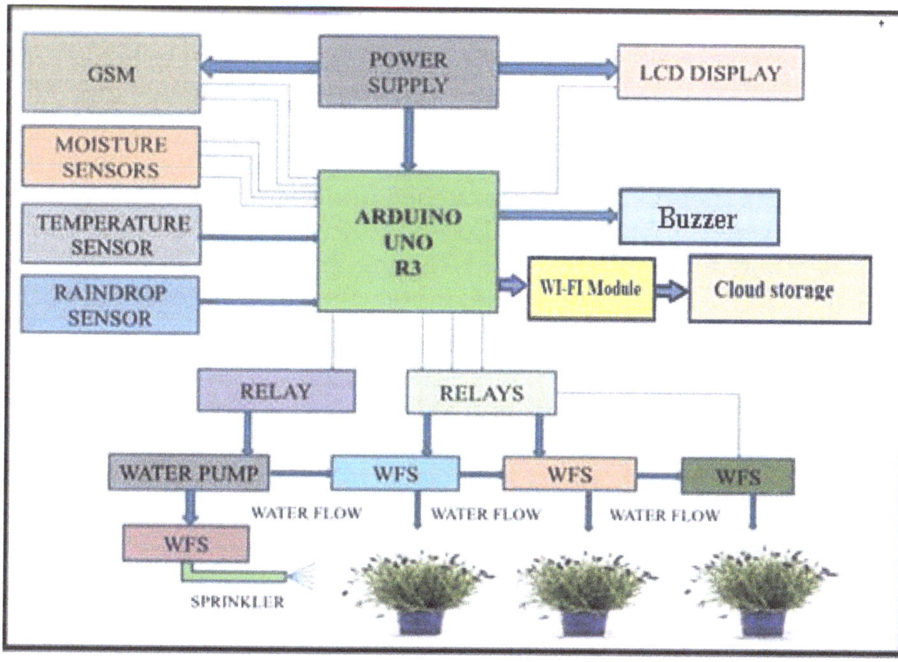

Fig. 4: IOT-based Raise Farming complete Structure

RESEARCH

While neediness planning joined with crop climate arrangement frameworks gives the instruments to target crop research and permit distinguishing proof and evaluation of a portion of the wide agro-biological and financial limitations that the unfortunate face in the minimal conditions where they reside, extra apparatuses that get the neighborhood settings which the poor live and cultivate are expected to make research pertinent and suitable.

The neighborhood settings of poor people, how they see themselves, their concerns and limitations, also as any open doors must be brought into the front of the harvest research process in the event that they take part in the exploration prioritization, advancement and execution.

CONCLUSIONS

Creating and doing trim exploration that benefits unfortunate ranchers in minor region of the it is complicated and challenging to foster world. It doesn't just involve applying the examination approaches and techniques that have been so effective for ranchers in preferred regions, however growing new ones that answer the specific states of these ranchers and regions. This implies explicitly focusing on them and tending to their requirements and conditions.

It requires incorporating approaches and techniques from various disciplines, yet from fields inside these that typically are not remembered to go together, for example, stress rearing and non-market valuation strategies; as well as nearer and more efficient connections among researchers and ranchers.

At long last, for crop examination to have a wide effect, it is important to work and connection various scales, from the worldwide to the neighborhood as well as the other way around. Every one of these require solid specialized and logical abilities as well as a promise to make research that is designated, applicable and proper for these ranchers, their families and networks.

REFERENCES

[1] Evenson RE, Gollin D. Assessing the impact of the Green Revolution 1960 to 2000. Science 2022;300(5620):758–62.

[2] Pingali PL. Milestones in Impact Assessment Research in the CGIAR, 1970–1999, With an Annotated Bibliography of Impact Assessment Studies Conducted in the CGIAR, 1970–1999,. Prepared by Feldmann MP. Standing Panel on Impact.

[3] Alston JM, Norton GW, Pardey PG. Science under Scarcity: Principles and Practice for Agricultural Research Evaluation and Priority Setting. Cornell University Press, Ithaca, NY; 2020.

[4] Renkow M. Poverty, productivity and production environment: a review of the evidence. Food Policy 2020;25(4):463–78.

[5] Kerr J, Kovalli S. Impact of agricultural research on poverty alleviation: conceptual framework with illustrations from the literature. Environment and Production Technology Division Discussion Paper 58. International Food Policy Research Institute, Washington, DC; 2019.

[6] Fan S, Chan-Kang C. Returns to investment in less-favored areas in developing countries: a synthesis of evidence and implications for Africa. Food Policy 2014;29(4):431–44.

[7] Almekinders CJM, Elings A. Collaboration of farmers and breeders: participatory crop improvement in perspective. Euphytica 2019;122(3):425–38.

CITATIONS

- Jain, A., Kumar, A. and Sharma, S., 2015. Comparative Design and Analysis of Mesh, Torus and Ring NoC. Procedia Computer Science, 48, pp.330-337.
- Jain, A., Dwivedi, R., Kumar, A. and Sharma, S., 2017. Scalable design and synthesis of 3D mesh network on chip. In Proceeding of International Conference on Intelligent Communication, Control and Devices (pp. 661-666). Springer, Singapore.
- Jain, A., Gahlot, A.K., Dwivedi, R., Kumar, A. and Sharma, S.K., 2018. Fat Tree NoC Design and Synthesis. In Intelligent Communication, Control and Devices (pp. 1749-1756). Springer, Singapore.
- Jain, A., Dwivedi, R., Kumar, A. and Sharma, S., 2017. Scalable design and synthesis of 3D mesh network on chip. In Proceeding of International Conference on Intelligent Communication, Control and Devices (pp. 661-666). Springer, Singapore.
- Sharma, S.K., Jain, A., Gupta, K., Prasad, D. and Singh, V., 2019. An internal schematic view and simulation of major diagonal mesh network-on-chip. Journal of Computational and Theoretical Nanoscience, 16(10), pp.4412-4417.